D1694431

Ideas

GERD FOLKERS AND
MARTIN SCHMID (EDS.)

IdEAS

ETH Zurich: Where Context Favors Innovation

CHRONOS

ETH
Eidgenössische Technische Hochschule Zürich
Swiss Federal Institute of Technology Zurich

COLLEGIUM HELVETICUM

Concept and Design: Gerd Folkers, Martin Schmid, Jürg Brunnschweiler

Organization, Editing and Proofreading: Martin Schmid

Translation: Angela Harp

Illustrations: Lorenz Meier

Layout: Thea Sautter

Printing, Finishing: Freiburger graphische Betriebe, Freiburg

© 2014 Chronos Verlag, Zurich, Collegium Helveticum and ETH Zurich
ISBN 978-3-0340-1242-3

7	**Foreword** GERD FOLKERS / MARTIN SCHMID
11	**Hotbed of Technological Change** RALPH EICHLER
23	**Molecular Megalomania** ULRICH W. SUTER
33	**This Is not How Physics Is Done! Or Is It?** GIOVANNI FELDER
45	**Talents, Education and Freedom— The Engine of Innovation** ROLAND SIEGWART
57	**There Is Nothing so Practical as a Good Theory** LINO GUZZELLA
63	**Janos, Johann, John—Budapest, Zurich, Princeton** MICHAEL AMBÜHL
73	**Seeking Sustainability** PETER EDWARDS / GERHARD SCHMITT
85	**When Photography and Microscopes Fail** BEAT H. MEIER
93	**An Eternal Vision** GERD FOLKERS / MARTIN SCHMID
105	**Simpler Living in the 2000-Watt Society** OLAF KÜBLER
117	**The Fascination over New York: "Les gratte-ciel sont plus grands que les architects"** WERNER OECHSLIN

Foreword

GERD FOLKERS AND MARTIN SCHMID

Scientific ideas are by no means tied to places, nor can they be pinned to a single point of origin, could this even be determined. Ideas are ephemeral and at most belong only to the mastermind who came up with them, or their associated research team. Considering that mobility and the global nature of scientific research is not a phenomenon unique to the 21st century, but one that has gone on for centuries, it is clearly not always easy to identify where exactly scientific ideas are born. Take Albert Einstein, for example; his career took him around the world. He made stops at ETH Zurich, the University of Bern, the University of Zurich, the University of Prague, then back to ETH Zurich, the Prussian Academy of Sciences in Berlin thereafter, and finally Princeton University. John von Neumann, to add another example, left his mark at the Friedrich-Willhelms-Universität in Berlin (now the Humboldt University), at ETH Zurich, the University of Budapest, the University of Göttingen, then back at Berlin, the University of Hamburg, ending up at Princeton as well. And yet: despite this mobility, the way they shaped and formed their ideas deeply depended on a particular environment. Scientific research cannot be conducted in a vacuum; it needs to take place in an atmosphere that inspires and offers the necessary organizational framework, infrastructure, financial resources, and so on—only then can it function as a hotbed ripe with scientific ideas and innovation. One such place is ETH Zurich; it has flourished in this role throughout its 150-year history.

An idea might thrive best in a greenhouse after it initially germinates, but once it is fully grown—published or patented—it becomes mobile. In this sense, ideas have always been of way of building bridges between nations, cultures and continents. It is within this context that this very book "Ideas" will be released at an event being held in New York entitled, "Zurich Meets New York". In May 2014, representatives of science, culture and business from both Zurich and New York will meet to discuss the innovations and visionary ideas that have come to life in Zurich. The gathering is being organized by the City of Zurich, the University of Zurich and ETH Zurich. This booklet takes a close look at ETH Zurich, shedding light on its current cutting-edge research and looking back at its history of innovation.

What is considered the biggest and best, most impressive, most sustainable idea ever developed at ETH Zurich? This is the question the editors posed to a select group of distinguished ETH professors who were each asked to respond with a short essay.

Ideas born at ETH Zurich? To give a characteristic overview of the ideas that emerged at ETH Zurich over the generations would take much more than the 100 pages of this brief volume. But that's not really the aim of this book. It is not an exhaustive list of every scientific achievement associated with ETH Zurich, or with its 21 Nobel laureates or the numerous other researchers who have been honored with the Turing Award, the Pritzker Prize and the Fields Medal, let alone all the other scientific prizes won by those linked to ETH. Rather, it is intended to be an illuminating glimpse inside the exciting, wide-ranging world of science at ETH Zurich. This small collection of stories aims to do one thing above all else — to take you on a fun scientific journey that will hopefully inspire you to seek out more stories of the like, or even better, to come up with your own ideas to be developed within the innovative walls of ETH Zurich.

The President of ETH Zurich describes in his introduction what ETH represents and how it has built up its reputation to provide an environment that encourages and drives innovation. Ten essays follow his introduction. Some reveal the difficulties in establishing innovative ideas or in being recognized for them; how identical ideas were born at two universities in parallel (at ETH Zurich and Cambridge) and how they only made their breakthrough at Zurich. Others tell how ETH Zurich has taken concepts such as sustainability or the 2000-Watt Society and contributed to their advancement.

A special thanks to all the authors who in addition to their daily responsibilities found the time to write an essay. A warm thank you, as well, to Lorenz Meier, who provided the impressive illustrations to complement and provide a visual counterpoint to the ideas described, and to Angela Harp, who translated the original German essays into English and adapted their style to suit an Anglo-Saxon readership.

Hotbed of Technological Change

RALPH EICHLER

Anyone who wants to understand ETH Zurich needs to take a deeper look at Switzerland and the ideology behind this small country with eight million inhabitants and four national languages. Located in the heart of Europe, Switzerland has no real natural resources, but it has built up a highly effective educational system and an excellent infrastructure, both of which are important to attracting top talent in science and business. Unlike crisis-riddled Europe, Switzerland enjoys a virtually low unemployment rate and material prosperity. This was certainly not the case when the Swiss Confederation was founded in 1848.

To develop what was then a relatively young state, Zurich politician and business leader Alfred Escher modernized the country's transport systems and expanded the railways across the Alps, which required both investment capital and skilled engineers. Escher first founded the Polytechnikum with other colleagues in 1855, which eventually was renamed ETH, the *Eidgenössische Technische Hochschule* (Swiss Federal Institute of Technology) to acquire the necessary skill and know-how to fulfill his dream. In addition to having investments in mind,

Escher also needed backing from financial institutions. By establishing the Schweizerische Kreditanstalt (now Credit Suisse) in 1856 and the Schweizerischen Rückversicherungs-Anstalt (now Swiss Re) in 1863, Zurich became the leading banking and insurance center in Switzerland.

It was over a hundred years ago that engineers laid the foundation for today's prosperity, building railways, tunnels and power plants across the nation. The "vision of Switzerland" thus took shape through the professionals working and researching at ETH Zurich and their successful innovations. Major firms such as Brown, Boveri & Cie. (BBC, now Asea Brown Boveri, ABB), Gebrüder Sulzer (now Sulzer AG) and Escher, Wyss & Cie. (later Escher Wyss AG and then acquired by Sulzer AG in 1969) were built on the foundational expertise of Swiss engineering specialists who produced the machines and turbines that became so popular on the world market.

In the early days of the Polytechnikum, professors and a large portion of the student body were recruited from abroad, especially from Germany. 159 years later, ETH Zurich remains an international university, with increasing global talents. Two-thirds of the professors, 20 percent of the undergraduates, 38 percent of the master's degree students and 67 percent of the doctoral students come from other countries.

ETH Zurich is known for its Nobel Prize winners, including the likes of Wilhelm Conrad Röntgen, Albert Einstein and Wolfgang Pauli who all worked at the university at one time. But just as important are the pioneering technical achievements. These include the foundations of the common-rail injection system for internal combustion engines, the regulation of catalytic converters and CMOS transmitter chip technology used in today's mobile devices. And the list goes on.

Also unique at ETH Zurich is its Department of Architecture, where design and construction technology are given equal emphasis. This holistic approach gives the architectural product a special, unique quality in terms of both form and function. Moreover, ETH was an early advocate of sustainability concepts, forming the Alliance of Global Sustainability (AGS) together with MIT in Boston and Tokyo University. A few years ago, the Singapore-ETH Center for Sustainability was also established at the invitation of the National Research Foundation in Singapore. An international team of up to 200 experts conducts research on the cities of the future and addresses the issues of resilience in our complex and globally interconnected world.

I tend to describe the political system of Switzerland, a democracy built on consensus, in the jargon of my training as a physicist: the system is highly innovative, but also has a strong tendency towards moderation. That means: we turn ideas into reality only when they have been thought through. This pragmatic approach keeps us from making hasty decisions we might come to regret.

The Swiss also tend to delegate authority from above to leadership roles where the respective knowledge is located. As a result, ETH Zurich has a large degree of autonomy in its policies and administration. The university is overseen by the ETH Board, which also exercises strategic leadership. Its eleven members are representatives from science, business and politics, and are elected by the Swiss Federal Council, the joint executive of the Swiss Confederation.

The ETH Board presents ETH matters to the political authorities and the parliament. This Board is responsible for the ETH Domain, namely ETH Zurich, the EPFL (École polytechnique fédérale de Lausanne) and four additional research institutes. It assesses the budget requests of these six institutions and distributes federal funding to them accordingly. The ETH Board also proposes a president for ETH Zurich as well as for EPFL to the Federal Council, which elects them.

Essential to the structure of ETH Zurich is how it recruits its academic staff. Appointing professors, in particular, forms the university's core recruiting process. The president and his staff are involved in all levels of strategic planning and seek to appoint only the world's leading scientists as academic personnel. The process makes it possible to locate such candidates relatively quickly, evaluate their credentials and get them on board if possible. More than half of all appointed professors were actively recruited and encouraged to apply. Outside of the usual hiring process, direct appointments in line with the principles of our *target of opportunity* are also possible.

ETH Zurich is well known for its research and is valued for its excellent education. Its close relationships with industry are intentional and contribute significantly to innovation within Switzerland. Almost a third of the country's CEOs hold degrees from ETH Zurich. Students who graduate from a Swiss high school (with the Matura) have an open choice of admission to all twelve Swiss universities, including ETH and EPFL. Tuition fees are low, representing less than 10 percent of a student's budget. After the first year of study, students take a first round of exams to qualify for permanent admission to a degree program at ETH. The exams may be retaken once. Usually, about two-thirds of the first-year students jump this hurdle. This rate might not seem very high, but the system has proved efficient and virtually no one drops out after passing these qualifying exams. 98 percent of the bachelor's degree recipients stay on at ETH to complete a master's as well.

It is important to take a balanced approach in a country as small as Switzerland: the two Federal Institutes of Technology in Zurich and Lausanne are here to train both the top academic elite and also the large number of engineers and scientists for work in industry. Therefore, the professor-student ratio is not nearly as good as in Anglo-American private universities. Since the Swiss economy needs more ETH graduates than Switzerland can produce from its own population, ETH recruits outstanding talent from abroad in addition to taking in its graduates within Switzerland. The language of instruction is therefore English in all of our master and PhD courses.

Ever more noticeable is the university's increasing impact on the entrepreneurial spirit in Switzerland. In 2013, 24 new companies were established by ETH researchers. Overall, ETH spin-offs made a total of over CHF 80 million in investments in the past year. It turns out that young companies marketing ETH knowledge are more likely to survive; almost 90 percent remain on the market after five years. That is twice the average survival rate for all Swiss start-ups. ETH spin-offs also create jobs at an above-average rate. These include companies backed by venture capital that provide significantly higher return on investment than those without external funding.

ETH considers knowledge transfer from the university to industry an important task. The ETH sponsored Technopark plays an important role in this, paving the way for pioneering companies to innovate; it acts as an incubator as well as multiplier for advanced technology in many industries. The planned Dübendorf innovation park, to be located on an abandoned military airfield in the Zurich metro region, also promises to be a hub in the Swiss innovation network.

Switzerland subtly promotes talents without extensive awards or PR. This happens at all levels and is also reflected in the successful dual education system. Only 20 percent of the country's schoolchildren will make it to a college preparatory high school each year, which in turn is an admission ticket to any Swiss university upon graduation. Most adolescents instead complete an apprenticeship with optional training at a technical college and coursework related to their future profession. ETH Zurich also trains and provides opportunities to these apprentices. Master technicians, for example, are hired to support ETH's top-level research and are proud of their contribution to science and society.

ETH Zurich has evolved into a technical university, offering degrees in mathematics, the natural sciences, engineering and architecture built on the strong fundamentals of mathematics, physics and chemistry. What was formerly known as the Polytechnikum, has recently grown into one of the leading universities on the continent according to international university rankings.

The university's organizational structure is characterized by a combination of strong leadership and flat hierarchies, which give ETH the flexibility to react quickly from the bottom up to new scientific and social developments. On the other hand, top-down leadership is effective in implementing developing project plans. The research funds made available to each professor also provide a competitive advantage, making it possible for faculty to test high-risk research ideas with funds they manage. Applying for competitive grants from funding agencies occurs after the initial stages of testing. As a result, 40 percent of all ETH grant applications to the European Union are successfully granted; 60 percent to the Swiss National Science Foundation.

The traditional organization of a university categorized into various disciplines is posing a problem for the grand challenges of society. Today, many of the major research focus areas such as energy, climate, food and urban design are interdisciplinary in nature. ETH Zurich therefore has begun to think outside the departmental box to promote interdisciplinary research in so-called competence centers. Prominent examples are visible in the research areas that deal with health issues at all levels, from molecules to the single cell to whole organs. Finding holistic solutions requires a wide band of expertise in biotechnology and food sciences, medical technology as well as knowledge from the engineering departments.

As ETH Zurich does not have its own medical school, it works closely with the University of Zurich and its hospital in the Verbund Hochschulmedizin Zurich. Another area for interdisciplinary research is energy, where knowledge from the worlds of mechanical and electrical engineering and computer science come together to form the Energy Science Center. Quantum engineering is also based on the synergy between physics and electrical engineering.

Recently, ETH Zurich founded a privately funded institute for theoretical studies (ETH-ITS). Similar to the renowned Institute for Advanced Study at Princeton, ETH-ITS will invite the best mathematicians and theoreticians and offer them a platform to indulge in pure research, free from other academic obligations.

In addition to experiment and theory, supercomputer simulations make up a third research dimension. Meeting the requirements of this new field, ETH currently operates the fastest computer in Europe. This high-performance machine located at the National Supercomputing Centre (CSCS) in Lugano is the most energy efficient of its kind in the world. With this computer, bright minds are investing a lot of time and energy into improving algorithms that could lead to more detailed climate models, more elaborate fluid dynamics and sophisticated materials.

Offering more than just research and teaching in the conventional sense, ETH is also home to other centers of expertise that provide distinct services to society and the economy. The Swiss Economic Research Institute, for example, issues economic forecasts that garner a lot of attention. For 75 years, this institute has been analyzing the ups and downs of the Swiss economy unlike any other institution and provides economic analysis on demand from positive to negative forecasts on inflation, recession or even depression.

Also well known is the Swiss Seismological Service (SED), which celebrates its centenary this year. Since its establishment, the SED has collected data on more than 13,100 earthquakes around the globe. On the political front, the Center for Security Studies (CSS) is of particular interest for security expertise and assessments. Whenever a political crisis erupts or flares up in the world, CSS experts are highly sought-after sources of information for the media.

Not to be forgotten are the research institutes of the ETH Domain that create valuable synergies with ETH Zurich, thanks to their joint governance and shared personnel working in research and teaching. For example, the Paul Scherrer Institute (PSI) is home to large research facilities.

The world has changed greatly since ETH Zurich was founded, as have the challenges facing science and technology. While Switzerland was once designing buildings out of concrete and steel, it is now developing virtual methods and tiny structures that have the potential to reshape the world, at least in the global North. Developments such as quantum computing, bio-based materials, and nanotech light particles or risk analysis results for financial systems, social and political-military conflicts, are all examples of the many progressive contributions stemming out of ETH research.

The impact of human activity on society, the economy and the environment has long crossed national borders. The "vision of Switzerland", still very provincial back in the early days of the Polytechnikum, has now extended its reach globally. Swiss universities are affected by globalization too as they face increased global competition for talent. Change will march on and continue to challenge ETH Zurich; but this "idea machine" in the Limmat city has been tackling the demands of the future for more than a century and it will not stop now.

Molecular Megalomania

ULRICH W. SUTER

Leucippus and his pupil Democritos had already arrived at the idea, around 2400 years ago, that matter could consist of atoms. According to these two philosophers, substances are heterogeneous at heart and are made of small, indivisible atoms and empty space. This philosophical curiosity eventually solidified into an accepted axiom among natural philosophers and forms the basis of how matter is understood today.

It was not until much later, in the 17th century, that the idea of the molecule was born: Atoms get together according to well-defined "architectural" plans to form stable assemblies. In 1661, Robert Boyle described this approach in his paper *The sceptical chymist*, in which he introduced terms such as "chemical anatomies" or "parcels of matter". His concept proved to be very successful, and laid the foundation for the emergence of modern chemistry. Many researchers developed the concept further until it had become common knowledge by the end of the 18th century. However, the mass (weight) of these molecules could only be determined at first in volatile substances in the gas phase, which is why the belief was widespread that all molecules are small—an idea that gradually established itself.

1861

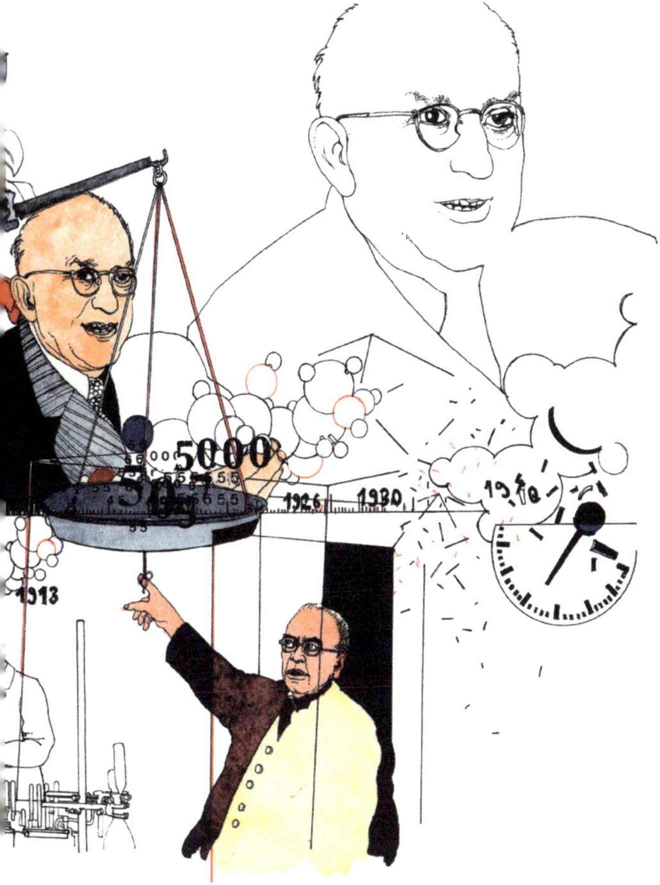
5000
1926
1930
1913

Although large molecules such as albumin and gelatin had been isolated and characterized even before 1800, their molecular weights could not be determined. In the beginning of the 19th century, the first experimental evidence of molecules with high weight materialized: Johann Friedrich Engelhard and Gerardus Johannes Mulder reported on various hemoglobins and proteins such as fibrin and egg and serum albumin, all with proven molecular weights in the tens of thousands. However, no credit was given to the results, as the discrepancy of the accepted values of volatile substances was too great.

The mystery of the "too high" molecular weights was apparently solved in 1861 by Scottish scientist Thomas Graham: He created the theory of the colloidal state—a kind of fourth form of matter, in addition to gas, solid and liquid. Graham concluded from various experiments with albumin and other natural substances in solution, that these existed as large aggregates (collections) of small molecules, as "colloids". This interpretation was strengthened by further observations over time, resulting in the predominant opinion that "secondary valence forces" between the small molecules would hold the aggregates together. This theory explained the confusing results that were obtained by François-Marie Raoult and Jacobus Henricus Hoff around 1880 using new methods of molecular weight determination. Substances with high molecular weight showed extremely unorthodox, highly non-ideal results in the physiochemical measurements.

This apparent success set the path for the great careers of two generations of scientists. The most famous names behind the colloid theory, which reached its peak between 1900 and 1925, were Carl Wilhelm Wolfgang Ostwald and Wolfgang Josef Pauli (father of the ETH professor and Nobel Prize winner for physics, Wolfgang Pauli). Their main conviction was that high-molecular weight substances are aggregates of small molecules. Finally, crystallography also came to the help of the supporters of the colloid theory: In analyzing crystalline macromolecular substances (for example, fibers made of cellulose or of stretched rubber), very small crystallographic unit cells were found. The colloid approach seemed to be right according to a crystallographic theorem, which stated that a molecule could not be larger than its unit cell.

In parallel to the colloid theory, another largely experiment-based development started in a different direction. The first scientist to synthesize materials with high molecular weight, analyze and calculate credible values for the molecular weights was Agostinho Vicente Lourenço, a Portuguese doctoral student in Paris. He managed to produce polymers from ethyleneglycol and succinates in 1860 and provided what is currently considered an accurate interpretation. Moreover, he suggested the possibility of copolymers. Heinrich Hlasiwetz and Josef Habermann adapted the ideas of Lourenço in 1871 to proteins, which they understood as condensates of molecular fragments, and after 1880 several other researchers found natural substances with molecular weights greater than 10,000, such as rubber and soluble derivatives of carbohydrates. Alfred Werner even postulated in 1896 that the Green Magnus salt ($Pt[NH_3]_4 PtCl_4$) contained long platinum chains. Although more and more researchers found results that constrasted with the colloid theory and supported Lourenço's explanation, their results were ignored — the collected results of these scientists were not enough to overturn the opinion of the scientific community as a whole.

By 1920, chemists (but also physicists and biologists) still commonly held the belief that there was a physical limit for molecular mass, which was surprisingly low from today's perspective. Molecules, so their creed, could not be big. They were equally convinced that pure substances must consist of a single molecular species. The greatest molecular weight was estimated in 1913 by Hermann Emil Fischer, the winner of the Nobel Prize in Chemistry in 1902: In a lecture at the Karlovy Vary Conference of German Natural Scientists and Physicians, he highlighted that high molecular weights were not possible, and pointed to the then largest known synthetic carbohydrate with a molecular weight of 4021. Paul Karrer, Nobel Laureate in Chemistry in 1937, also questioned those who believed in high molecular weights, and wrote in an article about the constitution of starch: "One has to wonder justly why the view that dozens or hundreds of glucose molecules are glucosidically assembled into long rows while forming starch could keep almost unshaken for decades." Karrer went on to explain why nature would certainly not do something like that for energy-related reasons.

The colloid theory had become a dogma. Apostates were ridiculed as ignorant. In a letter written by Nobel Prize laureate of 1927, Heinrich Wieland, the following can be read: "Dear colleague, give up the idea of large molecules; organic molecules with a molecular weight above 5000 do not exist!"

What would it take for a new view to prevail? A highly qualified, bull-necked fighter unwaveringly convinced by the idea of very large molecules. Herrmann Staudinger was this hero. Coming from the Technical University of Karlsruhe, he was appointed Professor of general chemistry (inorganic and organic chemistry) and head of the analytical laboratory at ETH Zurich in 1912. He found the idea of secondary valence forces abhorrent and began a series of detailed studies of high molecular weight substances; in particular, rubber, polystyrene (metastyrene) and polyoxymethylene (paraformaldehyde). In 1919, he declared his convictions for the first time in a lecture, asserting that polymers are "high molecular weight" substances, in whose molecules many atoms are chained together by strong bonding. In 1920, his paper *About polymerization* became a key publication of the field. He also applied the now common chemical structural formula in chain form—this alone was a snub to the entire scientific community. At this time his views were still theories, but in 1922 he delivered the first truly convincing experimental evidence—along with his collaborator Jacob Fritschi—in a paper on rubber and its hydrogenated derivative (a saturated hydrocarbon). Previously, the "colloidal" properties of the rubber were attributed to the many unsaturated double bonds in its molecular structure—now the hydrogenated rubber, which no longer contains such double bonds, behaves in the same manner as rubber. Staudinger and Fritschi also explained why such substances could not be composed of only one type (or length) of molecules.

Resistance among the ranks of the established sciences was enormous. The reports on his lectures—at that time, the lecture manuscripts, including the minutes of the subsequent discussion, were still published—point out the vehemence with which the discourse was held by both sides ("... does the professor not know that ..."). In 1926 Staudinger also coined the term "macromolecule", which was perceived by most members of the research community as a slap in the face, as a real insult. It took a firm and unbending person to survive the impact of "scientific expertise". One must assume that Staudinger led a hard life in Zurich.

After 14 years at ETH Zurich, Herrmann Staudinger moved to the University of Freiburg in 1926, where he received the opportunity to set up a great institution. In the following years, he steadfastly championed the molecular position that macromolecules with unlimited size existed. His research group systematically produced new polymer classes and over the coming years chemists, physicists and material scientists, and even biologists sided with him more and more — even if personal relations with him were not always easy. Staudinger was clearly a man of strong character and firm opinions.

In the beginning of the 1930^s, the fight for macromolecules was almost won. Staudinger was right and Lourenço's concept was accepted — finally! Only a few disciplines of science (e.g. biology) continued to hold fast to the colloid theory. Shortly after 1940, even those disciplines that had remained sceptical until the very end eventually adopted Staudinger's point of view.

What is fascinating about Staudinger's struggle for macromolecules is how an idea, although based on verifiable experimental facts and open for rational interpretations, can still take a century to be accepted by the scientific community, and half a century after it is accurately explained. Then finally, due to the power of reality that is undeniable in the long term, an inevitable breakthrough occurs. Science works!

This Is not How Physics Is Done! Or Is It?

GIOVANNI FELDER

"Ok Weyl, let it be. This is not how physics is done!" This is how Hermann Weyl, who formulated the principle of gauge invariance in 1918 at ETH Zurich, recalled Albert Einstein's reaction to his idea many years later. Was Einstein right? Yes and no. As Einstein rightly observed, Weyl's theory did not correspond to reality, even though it attempted to explain gravity, electricity and magnetism in one go in an elegant, consistent and mathematically stringent manner. On the other hand, Einstein's statement later contradicted itself: Today, gauge invariance is the fundamental principle that underlies the successful theory of the interactions between elementary particles.

The history of Weyl's idea began in 1913, just after the 28-year-old mathematician freshly completes his habilitation at the University of Göttingen and is appointed full professor at ETH Zurich. By the time he arrives in Zurich, the 34-year-old Albert Einstein has already been working there since 1912 as a professor of Theoretical Physics, and is far along in developing one of his greatest creations—the general theory of relativity, which explains gravity as a geometric property of spacetime. Ein-

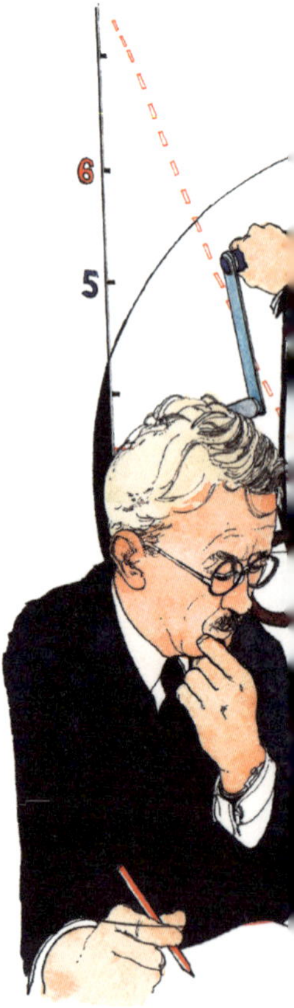
6
5

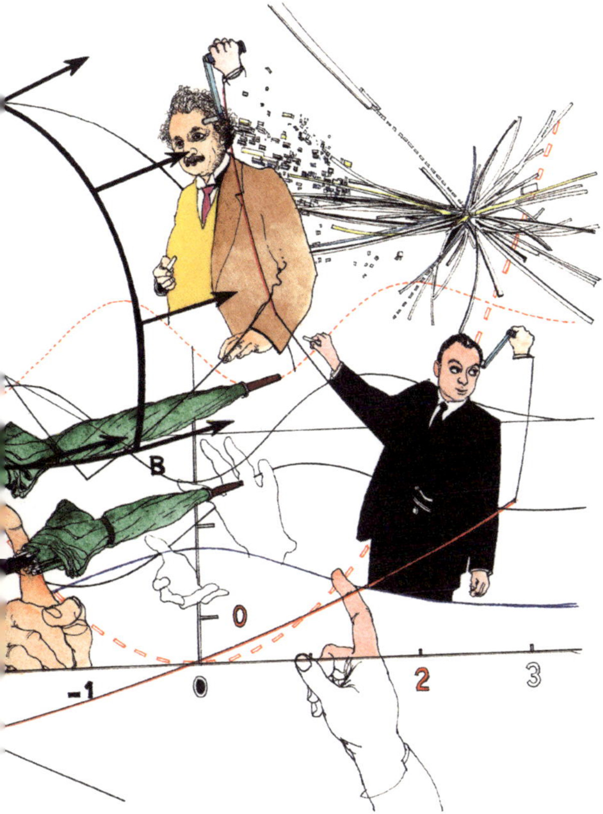
B
0
-1
0
2
3

stein and Weyl then spend the following academic year together at ETH Zurich before Einstein moves to Berlin where he — after a few failed attempts — presents his theory of gravity before the Prussian Academy of Sciences in 1915. During these years, Weyl is also occupied with Einstein's theory and the relationship between geometry and physics. Wanting to go a step further than Einstein, Weyl tries to geometrically explain electricity as Einstein did gravity. In pursuing this idea, he formulates the principle of gauge invariance.

What does gauge invariance mean? The origin of the notion is found in Einstein's theory of general relativity. Einstein postulated that space and time are curved under the influence of masses. Planets revolve around the sun instead of moving along straight lines, since space is curved by the mass of the sun like a billiard ball that is deflected on an uneven table. For Weyl, parallel transport, which was introduced by Tullio Levi-Civita in 1917, was crucial to interpreting curvature. Parallel transport can be illustrated by the following example: Take an umbrella or a walking stick and point it in a direction that we want to move in parallel. We then proceed in the pointed direction without changing the position of the umbrella. Then we turn to the left, keeping the umbrella in the same position, and cover a second distance. Finally, we turn to the left a second time — again without moving the umbrella — in order to return to our starting point along a straight line. That means we have moved along a triangle and moved an umbrella in parallel, so that it always pointed in the same direction. Unsurprisingly, the umbrella points at the end in the same direction as at start.

This experiment can be carried out rather harmlessly in your living room; but watch out for the crystal vase in the corner! However, it is advisable to only mentally visualize the following example: We now move our forward-facing umbrella along the equator to the east and proceed along a quarter of the circumference of the earth. Then we turn 90 degrees to the left to move, again without moving the umbrella, and go north. The umbrella still points eastward, namely to the right from our point of view, until we reach the North Pole. There, we make another 90-degree turn to the left and move forward—the direction of the umbrella is again unchanged—along a meridian to come back to the starting point. Now the umbrella points backwards, to the north—and not to the east as in the beginning of our trip. Thus, the direction has changed after the parallel transport, which is a manifestation of the curvature of the earth's surface. Einstein postulated that space and even time are curved: When we embark on a journey into space and point in a direction which is moved along in a parallel way, then the direction changes somewhat when returning, wherein the effect is small, as long as we stay far from strong gravitational forces. The direction, then, does not have any absolute significance in the general theory of relativity.

In this context Weyl's next thought is plausible: If directions change in parallel transport, then why not the lengths as well? Could it be that our umbrella moved in parallel has not only changed its direction on return from a journey into space, but also become a little shorter or a little longer? Weyl analyzed these questions and came up with the concept of gauge: The length of a line segment is only defined if we imagine putting a reference scale at each point with which we measure the length. Weyl referred to the totality of these imaginary references as gauge. The principle of gauge invariance states that the laws of physics, which are formulated geometrically,—namely they refer to the lengths and directions that are measured with respect to a gauge—are independent of the choice of the gauge. Weyl leads this principle further to its dynamical consequences: The gauge invariance is, as we have seen, closely connected with the concept of curvature, namely with the change of geometrical quantities by parallel transport. In Einstein's theory, the curvature has the physical significance of the gravitational field, which produces the gravitational force. In his extension of general relativity, Weyl identifies the curvature responsible for the change in length with the electromagnetic field, which mediates the forces on charged particles.

Einstein is skeptical from the beginning. Nevertheless, Weyl sends him his manuscript and asks him to present it before the Prussian Academy of Sciences in Berlin. This results in a lively correspondence between

Einstein and Weyl. On a postcard, Einstein writes: "Disregarding the fact that your theory disagrees with reality, it is certainly a great mental achievement." One of Einstein's main criticisms was that the clocks run slower or faster depending on their histories, a fact rather incongruous with the experiment. Weyl's 1918 paper finally appears in the proceedings of the Academy with an addendum from Einstein and a replication from Weyl, in which he points out the difficulty of interpreting time measurements.

In the midst of this controversial discussion, a 19-year-old student of physics at the Ludwig-Maximilians University in Munich makes his debut with a paper on Weyl's theory in 1919, shortly after graduating from the Döbinger high school in Vienna, in which the physical consequences of gauge invariance are published. The young Wolfgang Pauli was to become one of the most important physicists of the 20th century. Much later, in 1945, in a speech at Princeton in honor of Pauli's Nobel Prize, Weyl describes how the young Pauli approached his theory: "He dealt with it in a truly Paulinean fashion — namely with a pernicious blow."

As the debate settles, theoretical physics is rocked by a revolution that puts Weyl's idea in a completely new light: Quantum mechanics emerges around 1925, giving rise to a quantitative theory of quantum physics, in particular of the Bohr model of the atom. One of the main drivers of this shift was Erwin Schrödinger, a professor of theoretical physics at the University of Zurich since

1921. In his work entitled "On a Remarkable Property of Quantum Orbits of a Single Electron", written in 1922, he analyzes Weyl's theory in the context of quantum physics and makes a key observation: In the Bohr model of an atom, electrons move in atom-like planets around the sun, but only along very specific orbits. Schrödinger shows that the stretching factor of the lengths — predicted by Weyl — along these orbits is always an integral power of a basic factor. How big this basic factor is depends on an undetermined physical constant, which has the unit of measure of an action, namely the product of energy and time. Schrödinger speculated on the value of the physical constant. Combinations of known quantities such as the electron charge and the speed of light result in incredible stretching factors, numbers with hundreds of digits in their decimal form. Schrödinger plays with the idea that the sought physical constant corresponds to the square root of -1 times the Planck's constant, the physical constant of quantum physics. With this choice, the allowed Bohr electron orbits are precisely those in which lengths are not stretched after a revolution!

The fact that an imaginary number as the square root of -1 occurs in a physical law appears to be questionable at this time. Schrödinger writes: "I do not venture to decide whether this makes sense in the framework of Weyl's geometry of the world." It turns out that Schrödinger had guessed correctly. The framework to understand and realize this idea is created later with the Schrödinger equation (1925) that introduces an entirely new description of the physics at an atomic scale: Particles are no longer described as points that move along paths as in classical mechanics, but by wave functions that contains only the information about the probability of finding a particle at a certain place. What is new is that the gauge invariance is not related to the lengths of line segments but to the phase of the wave function. The wave function takes complex numbers as values, i.e. numbers, in which the square root of -1 is defined, so that Schrödinger's proposal of the physical constant is useful in determining the gauge factor.

This particular insight comes from the physicists Vladimir Fock and Schrödinger's student, Fritz London. The definite formulation appears after the development of quantum mechanics of the electron by Pauli and Paul Dirac, in Weyl's paper on "Electron and Gravitation" written in 1929 (his 1918 paper was published under the title "Gravitation and Electricity"). In this publication, Weyl significantly points out that gauge invariance and Einstein's general relativity theory explain the "origin and the necessity of the electromagnetic field". This singular idea was to serve as the basis for describing fundamental interactions in the coming decades: The symmetry principle of gauge invariance, according to which the laws of physics do not change when reference systems are changed independently at each point, forces the existence of a field, the gauge field, which mediates the interactions between particles. The gauge field, in the case of the electron analyzed by Weyl, is the electromagnetic field. Pauli, who in the meantime had become a professor of theoretical physics at the ETH Zurich, takes up this idea and applies it in his papers on wave fields in 1929 and 1930 in collaboration with Werner Heisenberg. It is here that the language of quantum field theory is introduced—a theory in which the principles of fundamental interactions is formulated today.

The way the idea of gauge invariance is later applied to the subatomic strong and weak interactions is interesting. In 1954, Chen Ning Yang and Robert Mills introduced the non-Abelian gauge theory, which is the relevant generalization of Weyl's theory. Pauli was severely critical of this, however. He had also written a manuscript with a similar result, but did not publish it because it seemed obviously wrong to him! Pauli argued that if the weak interaction, which is responsible for radioactive beta decay, would be described as in electromagnetism by a gauge theory there would be corresponding fields similar to the electromagnetic field and its quanta, the photons. But this was contrary to the experiment. It was not until 1964 that a possible solution to this objection was found in the work of Peter Higgs, Robert Brout and François Englert: If one postulates the existence of a particular field, now called the Higgs field, then the gauge fields of the weak interaction behave very differently from the electromagnetic field. Their quanta, the gauge bosons, then have a mass in contrast to the photon—thanks to this Higgs mechanism—and thus can only be produced and spotted in accelerators with sufficiently high energy. Today's accepted model for the description of fundamental interactions (with the exception of gravity) derives from the standard model designed in 1967 by Steven Weinberg and Abdus Salam. This successful model was based on a theory that was introduced by Sheldon Glashow in 1961; it is based on the principle of gauge invariance and also includes the Higgs mechanism as an essential building block. The gauge bosons that mediate the weak interaction were detected in the UA1 and UA2 experiments at CERN in 1983.

The conclusion of this story takes place almost 100 years after Weyl knocked on the door of Einstein's office at ETH Zurich. In 2012 the ATLAS and CMS research groups at the Large Hadron Collider at CERN discovered the Higgs particle, the quantum of the Higgs field. Thus, gauge invariance is confirmed as a fundamental principle of physics. Ironically gravity is not part of the standard model, although it was the starting point of Weyl's work: A theory of gravity compatible with quantum mechanics has yet to be found and is the subject of current research.

Literature:
For a more detailed description of these ideas and for references to the correspondence between Hermann Weyl and Albert Einstein see article by STRAUMANN, NORBERT (2006). *Gauge Principle and QED*. Acta Phys. Polon. B37 575-594 hep-ph/0509116.

Talents, Education and Freedom— The Engine of Innovation

ROLAND SIEGWART

According to many international rankings, the Innovation Union Scoreboard and the WEF Global Competitiveness Report among them, Switzerland is considered one of the most innovative and competitive countries in the world. Of course, this fact speaks to the success of Swiss innovation policy and notably what ETH Zurich contributes to the country's pioneering achievements.

The secret to success with regards to Swiss innovation policy is simple—there is no specific policy, nor do companies directly receive government aid. Rather, Switzerland promotes and maintains a strong enabling culture of freedom, moderate regulation and an excellent dual education system that supports talent in all areas, from carpentry to mechanics right through to engineering and astrophysics. The country also boasts excellent teaching and research institutions, ETH Zurich as one example, as well as first-rate public and private infrastructure and a stable, liberal economic environment.

Unlike most developed countries in the world, Switzerland has relied on a strong dual education system for decades with a comparatively low "matura" (baccalaure-

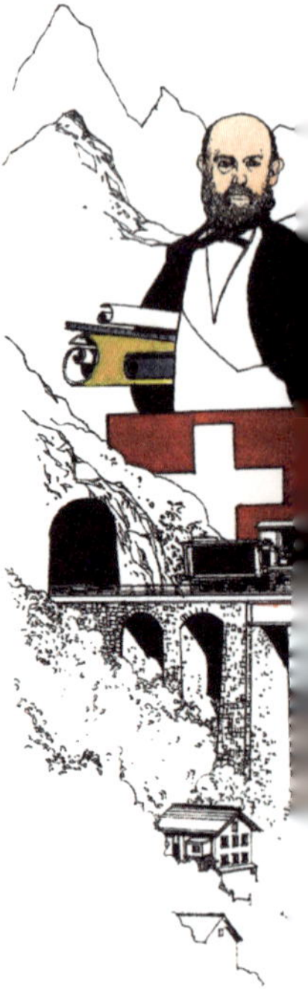

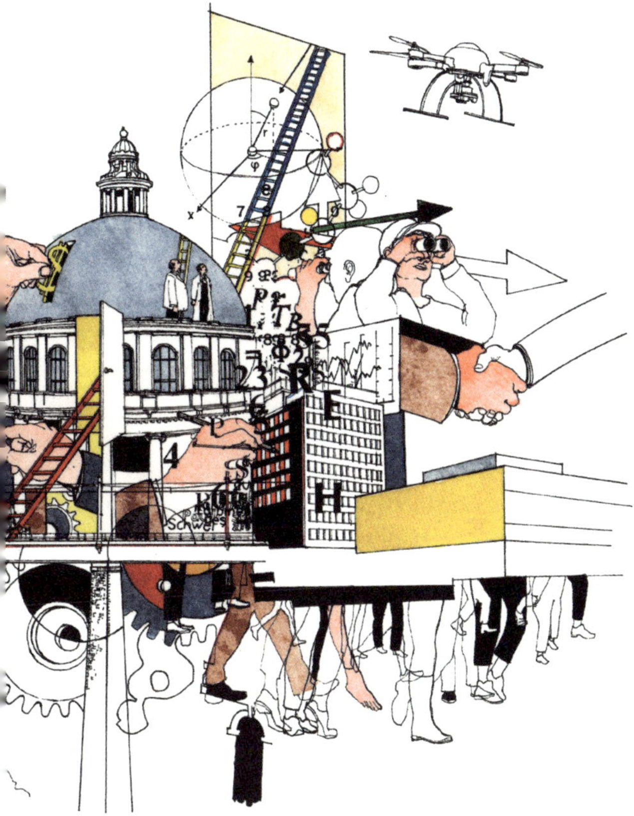

ate) quota and a correspondingly low rate of university graduates. Often the percentage of university graduates is seen as the quality benchmark of education and the educational level among the citizens of a country. But how is education actually defined? Does the percentage of adolescents graduating with a university degree really say something about the quality of education? Education is not simply an assortment of knowledge that can and should be drummed into the heads of our youths as homogenously as possible. Education is rather a teaching and learning environment that we offer our young people as a tool to finding the best possible way to develop specific skills and discover their talent. The Swiss education system is highly effective due to the fact that it allows young adults a variety of development paths, which ideally encourage progress. Contingent on one's assessment and interests, the education system provides academic and vocational training of the highest quality. This means that excellence in Switzerland, which in most countries is discussed only in the context of university education, is achieved in all areas covering all professions from the baker to the surgeon. As a result, the unemployment rate among juveniles is incomparably lower in Switzerland. Upon completion of vocational training, which combines practical on-the-job learning methods with classroom studies, graduates are highly qualified and professionally trained to immediately fit market needs.

Switzerland's education system is one of its greatest achievements and strengths by international comparison. High quality products arise not through prescribed quality management processes, but rather because employees at all levels have a clear understanding of their profession and carry out their work with enthusiasm and reliability, proactively contributing in the process.

Shortly after the Swiss federal state was institutionalized in 1848, Alfred Escher, pioneer, politician and the most influential thought leader during this time of change, was concerned about the development of Switzerland. In his speech at the opening of the national parliament meeting on 12 November 1849, Escher alluded to the risk of Switzerland becoming isolated, should it not emphatically expand the country's railway network to connect to Europe. Taking measures to avoid isolation, he proposed establishing a federal polytechnic and an institution for the purpose of financing the expansion of railroad infrastructure through the Gotthard. On this initiative, the young Swiss federal state founded the Eidgenössische Technische Hochschule (Swiss Federal Institute of Technology) in 1855, which is known today as ETH Zurich. The following year Escher constituted the Schweizerische Kreditanstalt, today's Credit Suisse. The estimated share capital determined by Credit Suisse in the amount of three million Swiss francs was exceeded 70 times within a few days by subscriptions of more than 218 million Swiss francs.

As the first technical university in Switzerland, the Eidgenössische Technische Hochschule should be the "bearer of the nation's future". The federal government's mandate to this new institution was, and is, to strengthen Switzerland's capacity for innovation, in addition to providing a first-rate education and research environment in technical and scientific fields. In the past 160 years, ETH Zurich has fulfilled this task with great success, constantly adapting to the rapidly changing needs of society. During its early developing stages, ETH Zurich starkly depended on the recruitment of professors from abroad, as there was simply a lack of competent researchers in Switzerland. Today, it still attracts professors from all over the world—an important factor of success.

Being a top university hinges directly and significantly on its ability to attract the best talent—topnotch professors, the most creative researchers and the most talented students. Outstanding scientists attract other outstanding scientists. Researchers and students of this caliber deserve academic freedom and the necessary infrastructure to fully flourish. ETH Zurich, therefore, fosters a culture of empowerment and trust coupled with flat hierarchies and efficient administration.

When it comes to research and education, ETH Zurich undoubtedly plays in the Champions League. But what about collaboration with the industry or knowledge and technology transfer, a third essential part of the federal government's mandate to ETH Zurich? By far, it is the university's graduates who contribute most to knowledge and technology transfer, as they have been trained, thanks to the close link between research and teaching, in the latest scientific findings. The knowledge they obtain is then taken and implemented in their professional environments. ETH Zurich breeds progressive thinkers and leaders, preparing them for Switzerland's corporate sector. This is demonstrated not least by the fact that around 30 percent of all senior managers working at Swiss companies have studied at ETH Zurich.

Increased globalization has intensified international competition significantly in the last few decades both in business and academia. In addition, innovation cycles are getting shorter. This has pushed international universities to place stronger emphasis on basic research, while companies concentrate on developments as closely in line as possible with market conditions. Consequently, the gap in the innovation chain between fundamental research at universities and application-oriented development in companies has widened over the years. This revealed the issue that promising ideas and technologies from universities hardly made it to industry, and those that did succeed entered with considerable difficulties. In order to sustain the country's innovation capacity, this gap urgently had to be closed. ETH Zurich has responded by strengthening its support for spin-off companies and by establishing strategic partnerships with industry, acknowledging innovation as a central element of knowledge and technology transfer.

About 25 years ago, it was under debate at ETH and among politicians and members of society as to whether a university should encourage the creation of spin-offs in general. At that time there were no regulations or guidelines to deal with university spin-offs. Should academic institutions get involved or would it be better keep out of it to avoid potential conflicts of interest and reputational risks? Today, spin-off companies are an indicator of the quality of a university, and entrepreneurial thinking is an important part of university culture; this is certainly the case at ETH Zurich. Annually more than 20 spin-offs originate from ETH Zurich; companies such as Sensirion AG or the listed u-Blox AG have each created hundreds of jobs and deliver microelectronic sensors for application in mass-markets, including cars and mobile phones. With its Pioneer Fellowships program, ETH Zurich supports students and scientists developing promising technology and clever business ideas on the way to becoming entrepreneurs. Recently, ETH Zurich started to couple these future spin-off founders with high-profile entrepreneurs and market experts from its business network. ETH Zurich also set up the Innovation and Entrepreneurship Labs with the aim to create a stimulating environment for its innovators. These labs have blossomed within a short time span into a viable melting pot of innovation.

On top of successfully achieving technology transfer through spin-off companies, ETH Zurich has been working closely with its industry partners to develop new ways to foster open and dynamic collaboration. Technology transfer is no longer a one-way street from academia to industry, but rather an equal exchange whereby both parties are mutually inspired and new visions are jointly materialized to face major societal and global challenges. A perfect example of this are the so-called *Lablets*, small company research units set up near university campuses. The *Disney Research Center* in Zurich is one such *lablet* that was established in a small building on the ETH campus in 2009. Led by ETH professor Markus Gross, research at the Disney Research Center in Zurich is carried out in close collaboration with ETH in various research areas such as capture technologies, postproduction video processing for the film industry, human and facial animation for films and computer games as well as sensor technology and robotics. Thanks to the close proximity of the *lablet* and the optimal synergies, this research platform has made remarkable progress at ETH Zurich and Disney alike. Collectively, the publication output has increased substantially, more than 180 individual projects have been launched and several dozens of patents registered. Markus Gross also received the Technical Achievement Award of the Academy of Motion Picture Arts and Sciences. The Disney Research Center in Zurich currently has more than 60 employees — many of them doctoral students from ETH Zurich. And as a result of this growth a new premises operates right next to the ETH campus.

The Binnig and Rohrer Nanotechnology Center (BRNC) in Rüschlikon, Switzerland is another successful example of a collaborative partnership between ETH Zurich and industry. Through this joint investment with IBM, the biggest public-private partnership of its kind in Switzerland, a clean-room laboratory was built, enabling premier research at an international level. The center is named in honor of Gerd Binnig and Heinrich Rohrer, the two IBM scientists and Nobel laureates who invented the scanning tunneling microscope at the IBM Zurich Research Lab in 1981. BRNC offers not only the most modern equipment for microfabrication, but also a stimulating environment for mutual scientific exchange between ETH Zurich and its industry partners.

While open partnerships between ETH Zurich and industrial partners was still rare a few years ago, today the university maintains several strategic partnerships with industry in a large range of areas, including world food supply, risk management or sustainable energy supply. Such partnerships quicken the process of establishing strategic research areas through endowed professorships and interdisciplinary research projects and allow for open discussion in common *Partnership Councils* on new visions and approaches.

As a leading university with vision and the necessary pragmatism, ETH Zurich is engaged in addressing the rapidly changing needs of society, science and industry. More than ever, it is an international center for education, scientific research and entrepreneurship — a place in Switzerland for groundbreaking discoveries and innovation.

There Is Nothing so Practical as a Good Theory[1]

LINO GUZZELLA

Getting a grip on instabilities or outwitting the forces of nature seems—like dreaming we can fly—an archetypal desire. Who doesn't know such games and tests of courage in small and big children? Acrobats even make a career of it—engineers as well.

Stability is an ancient but central topic in the engineering sciences. Stabilizing a machine or a process to a point where safe and profitable operation is possible, despite the impact of constant error, is one of the core tasks of engineering. Solutions, such as simple systems that can be kept stable, have been common for quite some time. Complex systems, on the other hand, were non-existent until Aurel Stodola, while in Zurich, asked the right colleague the right question—a correct answer was given, too.

Aurel Stodola, born in a part of the former Austrian Empire, which belongs now to Slovakia, was appointed professor of mechanical engineering and machine design in 1892 at the Eidgenössische Technische Hochschule, now ETH Zurich. He taught and conducted research there in various fields, including in control engi-

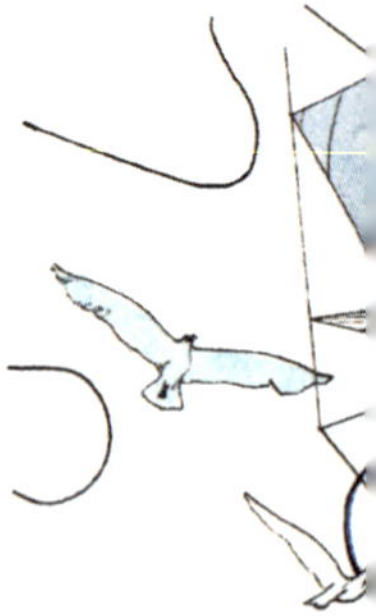

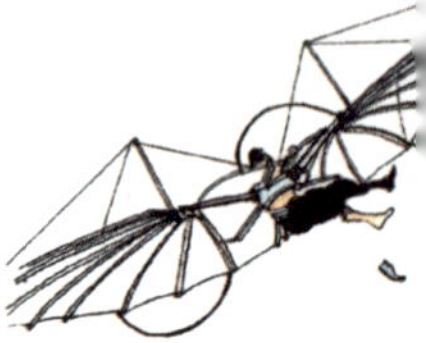

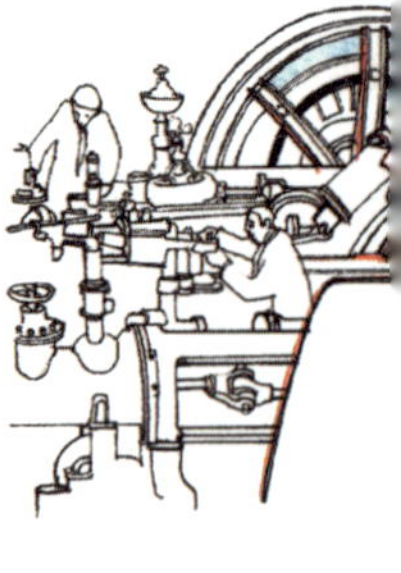

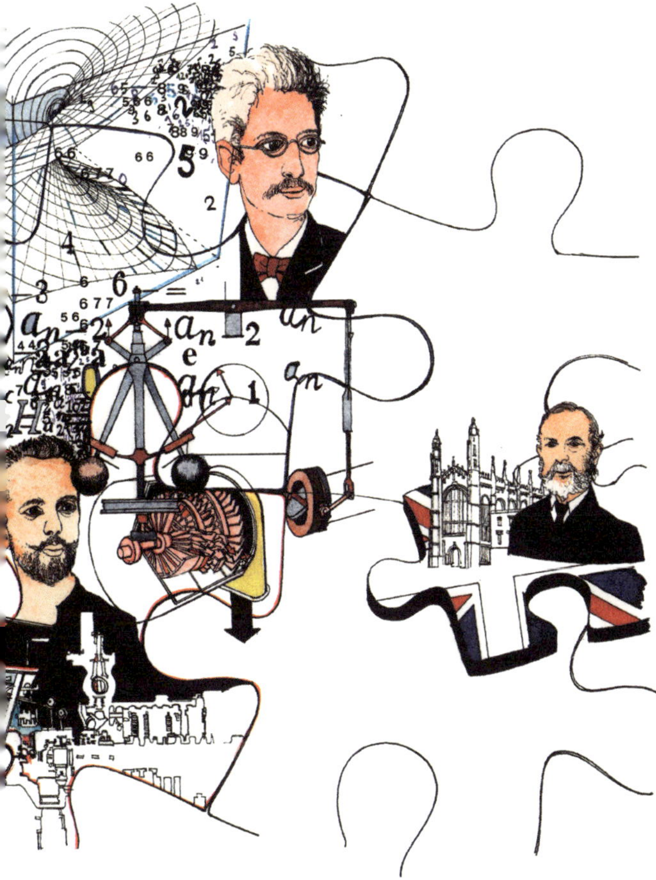

neering, until his retirement in 1929. In the first half of the 1890s, Stodola was occupied with a large theoretical study of water turbines that, among other things, was based on practical tests of experimental turbines. Eventually he succeeded in simplifying the mathematical modeling, but was not able to find a general solution to the problem of stability. During this research project, Stodola proposed the problem to Adolf Hurwitz, a mathematician appointed professor at ETH Zurich in the same year. Hurwitz, from Hildesheim, Germany, then managed to develop a mathematical method generally applicable to studying the instability of complex systems. Stodola, having been very pleased with this mathematical solution, included Hurwitz' discovery in his paper published in the *Schweizerische Bauzeitung*, a scientific journal for engineers. In turn, Hurwitz published the results of their fruitful collaboration in the *Mathematische Annalen* a year later, and wrote not without pride that Stodola had already applied the findings to a system at a turbine plant in Davos "with brilliant success".

Was this a groundbreaking scientific idea originally developed at the Swiss polytechnic? No! In Cambridge UK, the British mathematician Edward John Routh had developed the same mathematical procedure and published his idea about 25 years before Adolf Hurwitz. Routh's work on stability had probably remained unknown to the engineers on the main European continent due to the language barrier and the lack of the method's practical relevance. In 1898, Routh's book "The Dynamics of a System of Rigid Bodies", which contained his stability criterion, was translated into German. Stodola referred to it a year later in a paper. It was not until 1911 that the Italian mathematician Enrico Bompiani proved that the Routh and Hurwitz criterion were, in fact, identical. Since then, the Routh-Hurwitz Stability Criterion has become a standard application in control theory.

The birth of the Routh-Hurwitz Stability Criterion impressively illustrates how scientific research and the brilliant ideas it generates can emerge in various different ways and quickly find the path to practical implementation.

At Cambridge, a contest was the starting point for scientific work on the topic of "The Criterion of Dynamical Stability". The discovery of the stability criterion long remained known only to a small group of scientists. In Zurich, an engineer who worked on the construction of a turbine plant was enthusiastic about the solution. He even published the stability criterion—in an easy to use guide—in a magazine for engineers. It is in this manner that a solution to a practical problem is solved through theoretical findings. In turn, its practical application has become today's standard.

Still, the central aspect of this anecdote is rather that true breakthroughs and innovation often arise when boundaries are exceeded. Stodola was a brilliant engineer, yet his mathematical competencies fell short of what was needed to come up with the Hurtwitz-criterion. Conversely, Hurwitz would not have addressed the question of stability had not Stodolas coincidentally asked him. This ping-pong between the disciplines makes for an interesting approach. ETH Zurich has always strived to implement this approach—it is a principle of enablement that must continue in the future.

Literature
BISSELL, C. C. (1989). *Stodola, Hurwitz and the Genesis of the Stability Criterion. International Journal of Control*, 50:6, 2313-2332

Note
1 Quotation by Kurt Lewin (1890-1947), renowned German-American psychologist.

Janos, Johann, John— Budapest, Zurich, Princeton

MICHAEL AMBÜHL[1]

John von Neumann is considered one of the greatest mathematical geniuses of the 20th century. Contributing to a variety of fields, he was especially known for founding the game theory. In the 1920[s], he studied chemical engineering at ETH Zurich in Switzerland, and during this time he began developing the basic principles of game theory, principles that are still valid today. With this theory, strategic decision-making processes can be modeled in a mathematical way and used to analyze political and economical situations, particularly those where decision-making involves conflict or competition.

Game theory can be assigned to mathematics and the social sciences. Among other areas, it is useful, in analyzing international conflict (i.e. negotiations, arms embargo, military confrontations) or analyzing microeconomics (auctions, collective bargaining) or experimental research of conflict. Game theory has, in particular, influenced the field of social sciences. In fact, a number of scientists have received a Nobel Prize for their evolving work in game theory: Robert Aumann, John Harsanyi, Eric Maskin, Roger Myerson, John Nash, Reinhard

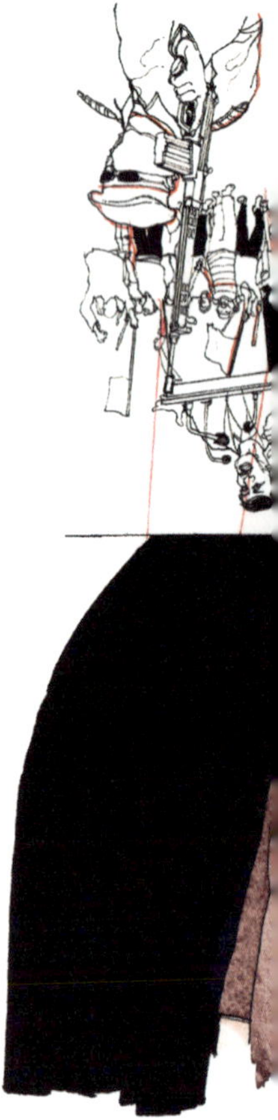

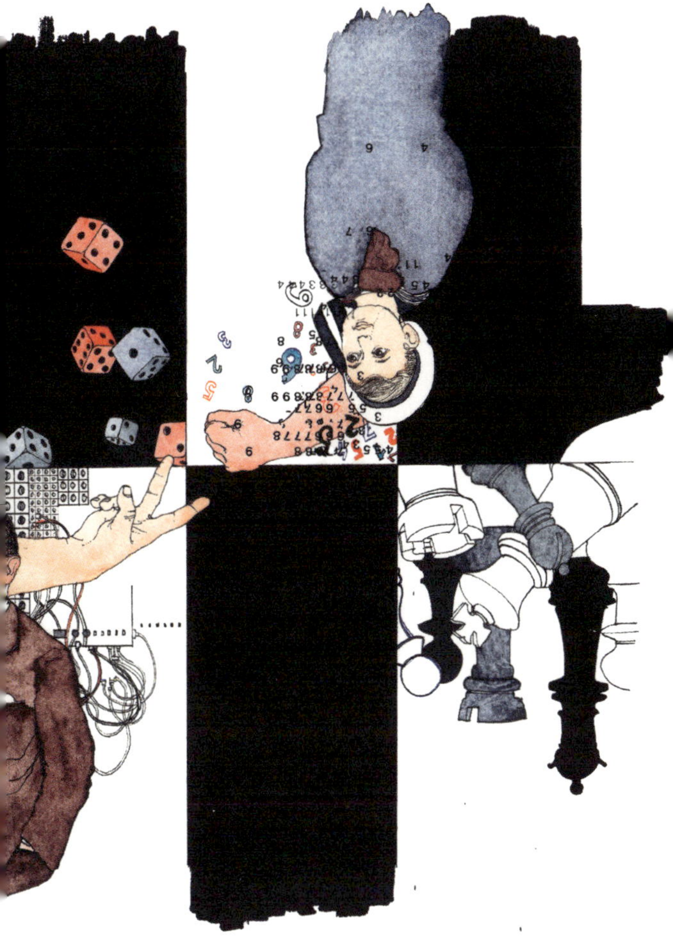

Selten, Lloyd Shapley; as well as George Akerlof, Leonid Hurwicz, Alvin Roth, Thomas Schelling, Herbert Simon, Joseph Stiglitz and William Vickrey for their achievements relating to game theory. John von Neumann himself unfortunately had no chance of such recognition, as he died 12 years before the Nobel Prize for economics first came into existence in 1969.

Von Neumann was born as Janos Neumann on 28 December 1903 into a family of bankers in Austria-Hungary, the former dual monarchy. As a child he already displayed an aptitude for mathematics. At the age of six he had the ability to quickly divide eight-digit numbers in his head, and his memory was exceptional: he could recite the contents of a book page after reading it. Subsequent to attending the German-speaking Lutheran Gymnasium in Budapest, von Neumann went on, following his father's advice, not to study "unprofitable mathematics", but rather chemistry. He first studied in Berlin, then in Zurich in October 1923 from the third semester on. Passing the obligatory entry exam with an average score of 6.0 (the highest possible score), he enrolled at ETH Zurich under the name of Johann von Neumann. Three years later, on 25 October 1926, he received his diploma in chemical engineering with a grade point average of 5.55.

Needless to say, his passion for mathematics never ceased during his chemical engineering studies. In addition to his chemistry courses, he would attend a number of math lectures, including those held by professor Hermann Weyl and George Polya, who like von Neumann was Hungarian-born. Professor Weyl's courses on the philosophy of mathematics and the theory of relativity were of particular interest to him. Von Neumann even taught Weyl's courses in his absence before having the academic credentials to do so. ETH professor Polya depicts von Neumann's incredible talent for mathematics in the following anecdote:

"The only student of mine I was ever intimidated by. He was so quick. There was a seminar for advanced students in Zurich that I was teaching and von Neumann was in the class. I came to a certain theorem, and I said it is not proofed and it may be difficult. Von Neumann did not say anything but after five minutes he raised his hand. When I called on him he went to the black- board and proceeded to write down the proof. After that I was afraid of von Neumann."

On 7 December 1926, six weeks after graduating from ETH Zurich, von Neumann presented a paper entitled "Zur Theorie der Gesellschaftsspiele" (On the Theory of Parlor Games) in a seminar at the Göttinger Mathematischen Gesellschaft. He sent the manuscript to *Mathematische Annalen* in July 1927, and it was published in 1928 (Vol 100, pp. 295-320). In his original 1926 paper, von Neumann basically only analyzed general societal games such as Roulette, Chess, Baccarat or Bridge, speculating how each of the players should play to achieve the best possible result. However, in theorizing this idea he raises, in his point of view, the core question of economics: "What will the utter most selfish 'homo economicus' do in the midst of certain external conditions?" He stated that, "any event with a given set of external conditions and given players [...] could also be regarded as a societal game when considering the effects it has on the people playing it". With this key thought he already bridges the gap to a theory that he will later extend to economics!

A game, von Neumann asserted, is composed of a series of events of which each can have at most a finite number of outcomes. In some situations, the outcome depends on coincidence, in others the will of the players. The aim of the players is to get the largest possible value of the [payoff] function f_m, in other words, the biggest, most desirable gain, whereby he postulated that $f_1 + f_2 + ... + f_n \equiv 0$. With this, von Neumann defines the zero-sum game, without yet naming it as such. According to this, the gains and losses of all players together is zero. In his paper, von Neumann also clarifies that he does not want to concentrate on probability games such as roulette, where the outcome is random, but rather on those in which the free will of the player is primarily crucial. The idea was to investigate the effects that players have on each other, considering the consequences of the fact that every player influences the results of all others, despite only being interested in their own. Von Neumann focuses on [non-cooperative] two-person [zero-sum] games, for which he proves Max Min = Min Max (*saddle point theorem*). A saddle point is a location where a maximum and a minimum come together — the highest point of a pass road, for example.

With the 1926 presentation in Göttingen, von Neumann laid down the first foundations for game theory. Where he exactly began developing this theory cannot be verified with absolute certainty due to the lack of precise documentation. However, it seems plausible that he formulated it while at ETH Zurich, where only six weeks earlier he had obtained a chemical engineering degree, and where he was intensely engaged in a scientific exchange with Weyl and Polya, two masterminds of mathematics. Even an ultra-quick thinker such as von Neumann was not likely to have invented this mathematical concept in a matter of a few days, nor could he have written a 30-page paper on it. It can therefore be assumed that the nucleus of the groundbreaking game theory emerged in Zurich: An idea born at ETH.

After completing his studies in Zurich, von Neumann worked at the universities of Göttingen, Berlin and Hamburg, as well as at Princeton University in New Jersey. In 1933, following Hitler's rise to power, von Neumann emigrated to the USA where he was appointed Chair of mathematics at Princeton. In 1936 he gave a highly acclaimed lecture on mathematical modeling of expanding economies. Together with the economist Oskar Morgenstern, John von Neumann (as he called himself after his emigration) eventually published the epochal work "The Theory of Games and Economic Behavior" in 1944, in which the important generalization is made on n-person games. The second edition, released in 1947, included the von Neumann-Morgenstern utility theorem and was to become a classic. It should be noted here that von Neumann not only wrote outstanding papers on game theory; he also made contributions, among many other things, to the axiomatic set theory, quantum mechanics, functional analysis and computer science (von Neumann architecture). In addition, he was a key figure in the Manhattan Project, acting as a consultant to the U.S. government in the planning of the first American atomic bomb and in the development of the military missile program. Among his scientific colleagues he was well respected for the way he generously passed on his ideas, and was furthermore considered a fun-loving and sociable co-worker with a special predilection for limericks and jokes.

The scientific seeds that John von Neumann successfully planted continued to bear fruit, especially in the second half of the 20th century where a real boom of new developments in the field of game theory surfaced. Among the many—just referring to the aforementioned list of Nobel Prize winners—John Nash, in particular, enhanced game theory by introducing the "Nash equilibrium" concept, a concept that was later named after him. He could prove the existence of this equilibrium for [finite] non-zero-sum games with more than two persons. In addition, Nash further developed the cooperative game theory and, based on the preliminary work of von Neumann and Morgenstern, formulated the "Nash bargaining solution". This solution is fair and efficient under the conditions of certain behavioral patterns (axioms) such as "pareto optimality", "not being worse off than in non-cooperative cases" and "symmetry".

Today, game theory is an analytical tool used in numerous applications: in economics (political economics and business administration), jurisprudence (economic analysis of law), political science, sociology, psychology, computer science and biology. It goes without saying that negotiation theory cannot be imagined without game theoretical analysis, as negotiations are by definition games in the sense of game theory, be it in the non-cooperative or in the cooperative form. It is for this reason that great importance is attached to game theoretical analysis within the newly created Chair of Negotiation and Conflict Management at ETH Zurich.

While game theory lends itself to the analytical acuity of negotiating situations, favoring thinking in models and serving as an important *indirect* method to solving strategic conflict, the *direct* benefits of the theory in specific negotiation circumstances has its limits—in a balanced assessment this cannot be left unmentioned. To obtain mathematical game theoretical-based assertions, one has to make model assumptions that simplify the reality. As a consequence, the results are sometimes not of concrete, direct relevance. However, it would indeed be unjust to the game theory to expect it to lead to practical solutions, for example, in the Syrian conflict or in the conciliation talks in the South Caucasus. Israeli game theorist Ariel Rubinstein gave a similar assessment in an interview in the *Frankfurter Allgemeine Zeitung* on 27 March 2013.

Game theory, originally founded by John von Neumann in Zurich, and later evolved with the help of many others, has progressively gained importance over the years, as it instructs clear and concept-based analytical thinking, from which valuable insights into decision-making situations emerge. Such situations are those for which there is no simple recipe to the solution.

Literature
ETH Zurich Archive, including course outlines and syllabi for the winter semesters of 1923/24, to SS 1926.
MACRAE, N. (1992). *John von Neumann: The Scientific Genius Who Pioneered the Modern Computer, Game Theory, Nuclear Deterrence, and much more*. New York: Pantheon Books.
POLYA, G. (1957). *How to Solve it; A new Aspect of Mathematical Method*. 2nd ed. New York: Princeton University Press.
VON NEUMANN, J. (1928). *Zur Theorie der Gesellschaftsspiele*. Mathematische Annalen, 100, 295-320

Note
1   Supported by Martin Schmid of Collegium Helveticum.

Seeking Sustainability

PETER EDWARDS AND GERHARD SCHMITT

The 1992 United Nations Conference on Environment and Development in Rio de Janeiro was a wake-up call to a world living beyond its means. Speaker after speaker drew attention to the planet's disappearing biodiversity, the damaged soils, the contaminated water supplies, and the warming climate. Many also presented grim statistics on widespread poverty and malnutrition, and high rates of illiteracy and infant mortality in many countries. The solution to these twin problems, it was argued, lay in finding new ways of maintaining economic growth that did not damage the environment for future generations. But how this could be achieved was far from clear. In its final declaration, therefore, the conference called for greatly increased research and capacity building aimed at meeting the challenges of sustainable development.

ETH Zurich was amongst the first major universities to respond to this call, making sustainability a strategic priority. This was reflected not only in its research and education, but also in its efforts to build and manage the campus sustainably, to support student initiatives, and to promote public understanding and debate. Together with other universities around the world, ETH Zurich has also

INNOVATION

taken a lead in promoting international research collaboration. En route, the Swiss university has not gone without facing challenges, but in launching two important initiatives and striving to carry them out it has, as a result, gained valuable insight and intellectual growth.

*

Soon after Rio, Jakob Nuesch, the ETH president at the time, conceived the idea that faster progress towards sustainability would be possible if the world's leading technological universities were to collaborate more closely. This idea found an immediate echo with the presidents of MIT and Tokyo University, and in 1994 the three universities agreed to form a partnership to be called the Alliance for Global Sustainability (AGS). A fourth partner, Chalmers University in Sweden, was to join a few years later. With generous financial support from the Swiss industrialist Stefan Schmidheiny, AGS soon developed an ambitious agenda of research and education based on three fundamental principles: (i) that sound science and technology must be brought to bear on environmental issues; (ii) that wise environmentalism is also good business practice; and (iii) that a global problem requires global cooperation for its solution.

AGS was guided by an international advisory board of industrialists and political leaders. Its distinguished members made it clear that AGS research should not merely be academic "business-as-usual", but must be aimed at delivering solutions. The rules for allocating research grants were made simple and non-bureaucratic: a strong focus on sustainability, and at least two AGS partners to be involved. Funding was competitive and applications were peer-reviewed, though existing peer-review procedures proved to be inadequate for this novel, problem-oriented type of research.

AGS soon had an impressive portfolio of research, on topics as diverse as the problem of arsenic in the water supply of Bangladesh villages, designing tools for sustainable buildings in developing countries, and understanding how businesses respond to different regulatory regimes intended to protect the environment.

AGS also pioneered a new type of large-scale collaborative project that involved external partners and sponsors from the earliest stages. One of these was the China Energy Technology Program, sponsored by the Swedish-Swiss industrial company ABB Ltd, which had the ambitious goal of helping China develop its electricity supply system in a way that was environmentally friendly, avoiding the mistakes that other industrializing nations had made. It was a hugely complex project, undertaken on a scale never previously been attempted. In all, it involved over 75 scientists from industry and from the partner universities, as well as many stakeholder groups in China.

Another large project was the Mexico City Case Study, in which scientists from many disciplines worked closely with city authorities and other stakeholders in an ambitious program aimed at understanding and defeating the city's chronic air pollution. Through a combination of research and outreach activities, the program was remarkably successful in developing systems to measure air quality and enable appropriate policy response. Indeed, within a few years its success could be measured directly in terms of improved health and reduced mortality amongst the citizens of Mexico City.

The AGS collaboration eventually came to a halt in 2011. By then, the topic of sustainability was firmly anchored in the teaching and research of all the partner institutions, and the approaches pioneered by AGS were being applied in new contexts and collaborations.

*

A new collaboration then emerged in Singapore. In 2006, Dr. Tony Tan, then Deputy Prime Minister (and now President) of Singapore, visited ETH Zurich and the ETH Domain to explore possibilities of cooperation. This visit led to a proposal for ETH to establish an independent research institution, the Singapore-ETH Centre for Global Environmental Sustainability (SEC) to undertake research in collaboration with Singaporean universities and government agencies. In early 2010, an agreement was signed and the Future Cities Laboratory (FCL), as the first program of SEC was initiated.

The creation of FCL was a direct response to the extraordinary growth of the urban population. Today, about 3.3 billion people live in cities, but this number could double by 2050. Not surprisingly, cities present some of the world's most intractable social, economic and environmental challenges. They consume a large fraction of all energy and natural resources, with potentially serious consequences for the local environment. But cities are also sites of diversity and creativity, center of innovation and entrepreneurship, and engines of economic activity. The central question posed by FCL, therefore, was how can cities be designed, built, managed and maintained to improve sustainability?

The FCL program was based on the concept of "urban metabolism", which treats the city as a complex system characterized by stocks and flows of resources, such as energy, water, capital, people and information. The individual FCL projects, led by professors at ETH and Singapore, cover topics as diverse as new building materials, digital fabrication, mobility and transportation planning, urban design, urban sociology and territorial planning. And like AGS, FCL also undertakes largely complex projects and intends to make an immediate difference to the urban environment. A good example is the "Cooler Calmer Singapore" project, which aims to increase the quality of life in Singapore by reducing heat and noise pollution in the city. Doing so requires a much deeper understanding about the anthropogenic sources of heat and noise, and which measures are likely to be most effective in reducing them. Scientists from many disciplines are collaborating with partners from government agencies, not only to understand the processes, but also to provide practical tools and guidelines that policymakers can use to make the city (and other cities like it) cooler, calmer and more liveable.

An important ingredient in the "glue" holding these projects together are the superb facilities provided by the "Value Lab Asia" for capturing and presenting 3-D and more-dimensional data. These facilities provide not only an essential research tool, but also a means for translating research into instruments that practitioners can use. For example, architects and planners can visualize the changing plumes of heat swirling around buildings as the wind changes, and explore how new designs might affect the city's heat balance. Or they can gain a bird's eye view of the city's traffic streaming through the streets, and test how new developments—adding a bus route, for example, or giving motorists real-time information about traffic congestion—would affect its flow. Not surprisingly, the Value Lab Asia attracts a steady stream of visitors from industry, government agencies and other research institutes.

In a remarkably short time FCL has established itself as a significant element in Singapore's research landscape, working in close partnership with Singaporean universities, with government agencies, and with industry. The city itself has proved an excellent place to study the problems and opportunities for sustainability posed by rapid urban growth, especially in tropical environments. For example, FCL researchers have shown how the efficiency of cooling systems can be dramatically improved, offering the prospect that future buildings will not only consume less energy but also require less materials to construct. And at a larger scale, FCL research has demonstrated how the complex inter-dependencies between cities and their surroundings can be characterized by flows of natural resources, people and wealth, and also by less tangible things such as ideas and cultural influences. These and other findings have obvious implications for the ways that cities are designed, built and managed, and are already attracting interest from policymakers.

*

Through these initiatives, ETH Zurich has pioneered new ways of doing research with the aim of providing practical answers to some of society's most pressing problems. But the road has not been an easy one. The participating scientists had to leave the relative comfort of their own discipline, where they were surrounded by like-minded colleagues and where their research was understood, and collaborate with people with different knowledge backgrounds and methods of approach. Two ETH colleagues, the environmental scientist Peter Baccini and the architect Franz Oswald, have vividly described the long, and sometimes difficult road that led to their successful collaboration on new types of urban design. One the one hand, the architects showed a "certain mistrust with regard to scientific methods", fearing that the reductionist approach did no justice to complex urban phenomena. On the other hand, the scientists and engineers, used to the relatively strict terminology of their disciplines, soon got lost in the bewildering diversity of metaphors used by the architects, whom they accused as "acting like chameleons".

However, mere interdisciplinary collaboration was not enough. In the early days of AGS, many academic researchers optimistically believed that they need only deliver a better solution—say a motor vehicle using less fuel—and society would adopt it. Their growing realization that this was not so was nicely captured in the title of an AGS workshop, "Why Bad Things Happen to Good Technology". Indeed, understanding the full complexity of problems, not only from a technical viewpoint but also from a societal one, was one of the reasons why both AGS and FCL found it important to develop larger projects that brought together many disciplines and relevant stakeholders. This in turn made great demands upon researchers in terms of better coordination and project management. As Baldur Eliasson of ABB Ltd wryly commented about the China energy project, "the process involved a lot of coordination, travel, communication, and patience, but ultimately was very rewarding". And these projects also posed complex questions for researchers:

how far should one depart from one's own field of expertise to meet the wishes of stakeholders? How much effort should be devoted to implementation as opposed to research? To what extent should one advocate particular solutions rather than remaining academically neutral? These are questions that we face every day, questions to which there are no easy answers.

But despite these difficulties, the two initiatives have been very rewarding, both for individual scientists and for the involved universities. Former MIT president Chuck Vest certainly spoke for all the AGS partners when he said that the formation of AGS had "released a wave of creative energy" that transformed his institution. Perhaps the greatest source of that energy was the students. From the beginning, students played a very active role in AGS, and used it as a launching pad for their own initiatives. For example, the climate protection partnership, MyClimate, began with an AGS meeting in Costa Rica, while the World Student Community for Sustainable Development was a spin-off from an AGS educational course. And there were similar student initiatives at partner universities. Furthermore, the doctoral students who receive their training within these programs benefited from working in a unique environment, one that developed their skills in communicating across disciplines, across sectors, and across cultures. Therefore, it is no surprise that many of the doctoral students in the AGS program have gone on to assume leading positions in academia, industry and administration. Although FCL is still in its infancy, a similar development has started for students of this program.

Essentially these initiatives have also met an important societal need. As the magnitude of the sustainability challenges becomes clearer, societies are looking to universities to take leadership, both in developing solutions and in providing a forum for discussion. "The challenges of sustainability are overwhelming"—"People will listen to academics"—"You are seen as impartial"—"Credible"—"Universities need to do more"—"Aggressive leadership is needed that only academia can provide"—These are the very messages being repeatedly heard among senior policy makers and leaders of industry.

*

As concern grows about the negative impact humans have on the environment, many argue that academic institutions should be doing more to help societies move towards sustainability. These critical voices point out that science has often been late or ineffective in preventing or mitigating emerging risks, so that even well-known environmental problems such as biodiversity loss and climate change remain unsolved. The new approaches pioneered by ETH Zurich and its partners offer one important way to begin bridging the gap between research and taking action.

When Photography and Microscopes Fail

BEAT H. MEIER

Creating images is of significant meaning in the human understanding of things and processes. Images can be concrete, like photographs, or they may only exist in our imagination. Particularly fascinating, are the images that draw our attention away from daily life and awaken our curiosity about the deep sea, space, the internal structure of living beings (e.g. cross-sections of a thinking brain), or images of objects that are far too small to be seen by the naked eye (e.g. biomolecules that make the brain work). The discovery and development of methods with which these objects and their movements can be made visible is an intriguing field in the natural sciences. When photography and microscopes fail because light cannot penetrate the skull or because the objects are significantly smaller than the wavelength of visible light (the proteins in the brain and body are typically 100 times smaller than the wavelength of visible light), innovative alternatives to microscopy become a necessity.

At ETH Zurich, research has focused particularly on inventing technologies that use radio waves with wavelengths in the centimeter range to construct images

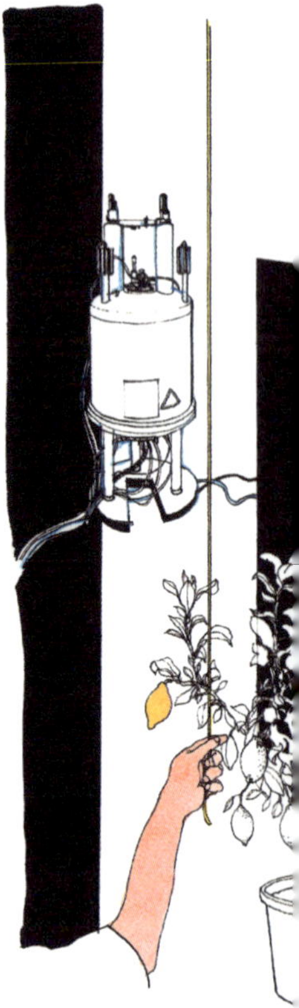

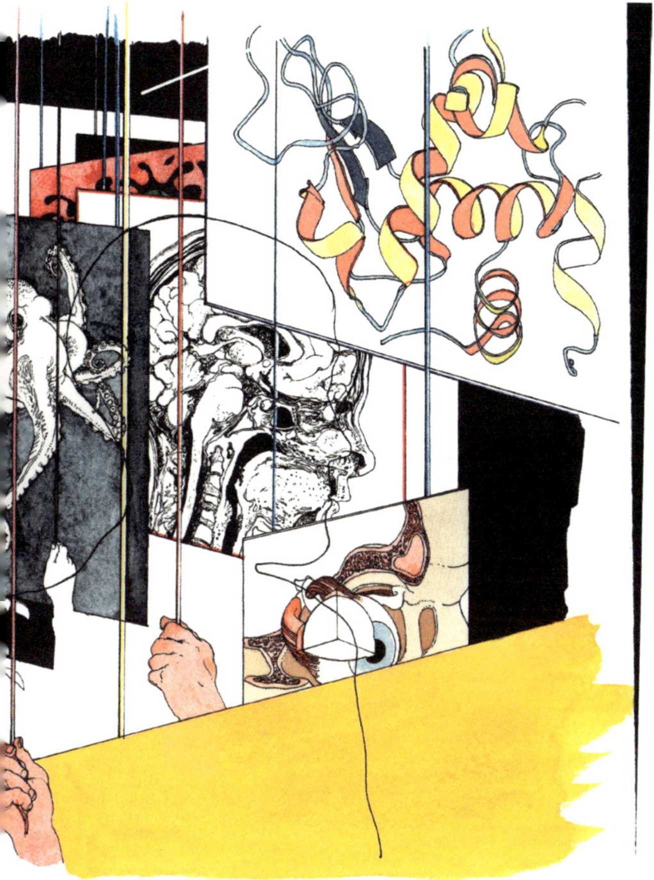

using spectroscopic methods. Spectroscopy deals with the interaction of electromagnetic radiation and small systems (atoms, molecules) and originally had relatively little to do with imaging. The precise wavelength of the absorbed or emitted radiation depends on how the direct environment of the radiating system looks. In our case, we discuss a somewhat abstract quantum mechanical property of the atomic nucleus known as nuclear spin. One can imagine a nuclear spin as a small gyroscope, the precession of which measures the strength of the magnetic field at a given location.

These nuclear spins were first experimentally detected after World War II by the former ETH student Felix Bloch, at that time a professor at Stanford and, independently, by Edward Purcell, who worked at Harvard University. It soon became clear that the precise absorption wavelength of the nuclear spin resonances was, in a complex way, dependent on its chemical environment and on the neighboring spins. The absorption or emission of electromagnetic waves by the spins, as a function of the wavelength or frequency, is called the spectrum. A biomolecule contains many spins in different environments. Therefore, due to the many wavelengths involved, the spectra of biomolecules were hopelessly complex.

The ETH professor Richard Ernst, inspired by the work of Jean Jeener, a Belgian scientist, and building on a long tradition of spectroscopy at the Laboratory for Physical Chemistry at ETH Zurich, discovered a way out of the complexity that just did not seem intuitive: Through more intricate experiments, the spectra of the nuclear spins dispersed into several dimensions. From one-dimensional lines that look like mountain silhouettes, arose mountains from two-dimensional surfaces. And thus, two-dimensional spectroscopy was born! It was later developed into three-, four-, and higher-dimensional spectroscopy. These concepts have influenced many other fields in chemistry and physics and have made it possible to not only characterize the properties of the spins and classify them based on one characteristic to a certain extent as in common spectroscopy, but also to represent the relationships amongst the spins. If conventional, one-dimensional spectroscopy is able to classify a group of humans, formulated in a simplified and model-like manner, based on a single characteristic, for instance size, two-dimensional spectroscopy allows for the representation of the relationships among the people—a much more interesting study, revealing who interacts with whom or who is close to whom.

The multi-dimensional spectroscopy was a wonderful invention, but, initially, it was merely an exciting toy for a handful of specialists. The biological background of ETH researcher Kurt Wüthrich and his close collaboration with Richard Ernst were necessary to create an instrument for determining structures of biological molecules which, in turn, led to breakthroughs in pharmacology and biology being at the origin of products with a direct benefit to mankind, for instance, new drugs.

Professors Ernst and Wüthrich were each awarded the Nobel Prize for Chemistry (1991 and 2002). The path from a scientific discovery to an applicable, marketable product often takes many years. The more fundamental the discovery is, the more continuing research is required to make the product "usable"; usable in the sense that it can be utilized by a broader group of people who have no specific interest in the precise details of the technology.

The fundamental and theoretical developments in two- and multi-dimensional spectroscopy also greatly influenced another field: magnetic resonance imaging, or MRI. This method, which already existed at the time of Ernst's groundbreaking work, was revolutionized thanks to the ideas of multi-dimensional spectroscopy, and is one of the most important imaging methods in the medical field today. It produces cross-section images of the human body, enriched with amazing additional information: Not only is the brain depicted, but one can also identify the areas of the brain that are active while we perform a certain activity such as reading, watching, speaking, thinking. This development also took many years. As a young doctoral student, the writer pondered about the several simple cables, ones similar to those in model railroads, that ran out from an experimental nuclear spin spectrometer and was told that these were the electrical connections used to record the first Fourier MRI image: an image of two capillaries filled with water. A long optimization phase followed using tolerant objects such as citrus fruits before the method began its triumphant tour through the hospitals of the world.

In contrast to more practically oriented inventions, which often lead to an applicable product within a short period of time, inventions in fundamental research require a longer start-up phase and have a lower level of recognition as far as the final products are concerned. However, they tend to have a more widespread impact on many individual fields. The discovery of multi-dimensional Fourier spectroscopies has led to capturing images of the brain and proteins, a deeper understanding of life processes, better medications, psychological and neurological realizations and improved plastics and intelligent materials. Even 35 years after their discovery, they continue to make an impact with an undiminished vigor!

An Eternal Vision

GERD FOLKERS AND MARTIN SCHMID

Do you know who Franz Urs Balthasar is? Are you familiar with the names Philipp Albert Stapfer, Stephano Franscini, Ignaz Paul Vital Troxler or Kasimir Pfyffer? Save the experts or those historically versed, hardly anyone knows of them. The persons listed were not graduates of ETH Zurich, nor did they never ever teach there. They were not even scientists, and they certainly did not make a name for themselves as founders of groundbreaking scientific ideas. Rather, all of them were politicians of the 18th century through to the first half of the 19th century, at a time long before ETH Zurich was founded. And yet, the personalities mentioned are in one way or another related to the university. Despite living in different time periods and having different origins and political orientations, they all had one thing in common: a vision to create a federal university! All of these men have long since passed away, their nearly names forgotten—yet, their vision, through ETH Zurich, is alive and flourishing.

*

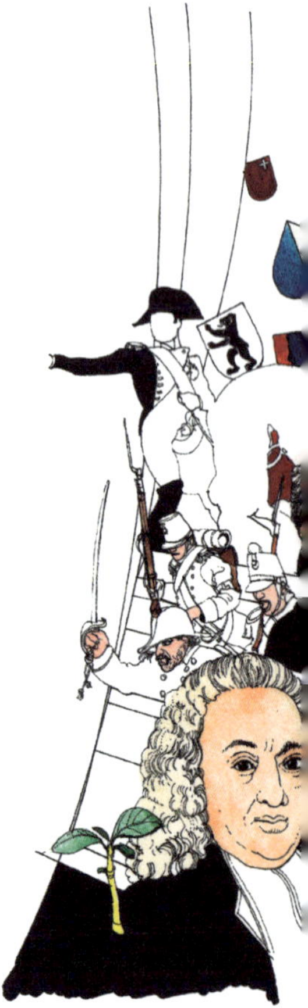

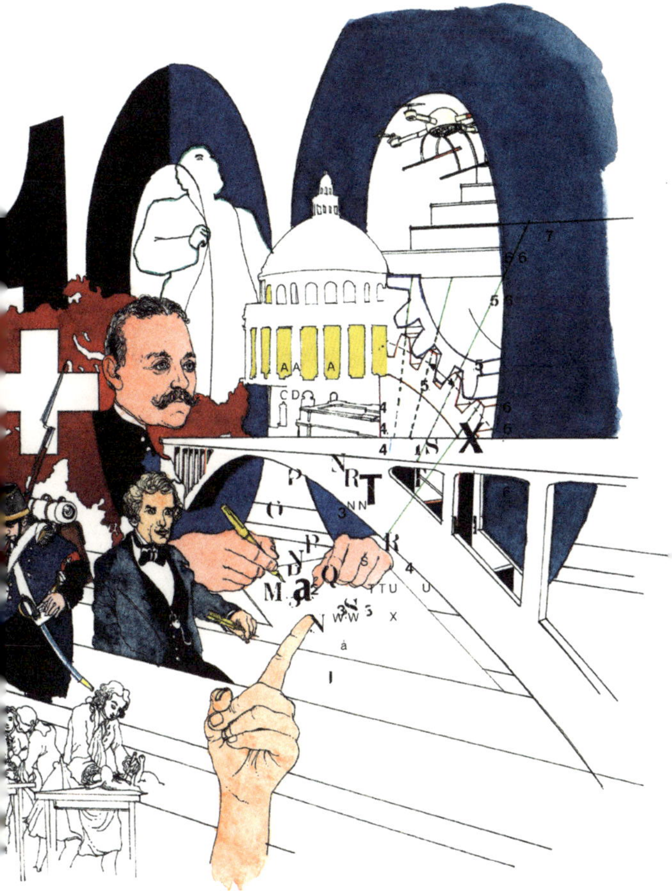

The idea of establishing a central Swiss educational institution came up for discussion among a wider political public for the first time in the mid-18th century. Franz Urs Balthasar (1689-1763), a lawyer and politician from Lucerne, brought the topic to light in his 1758 publication "Patriot. Träume eines Eydgenossen von einem Mittel, die veraltete Eydgenossenschaft wieder zu verjüngeren". Expressing a patriotic dream to ameliorate the Swiss Confederation, this work was broadly received in reform circles, particularly the novel idea to establish a federal institution of higher education that would serve as a "hothouse" for aspiring politicians of both denominations. His vision was that only law and political science should be taught—an idea still far away from the modern university concept. Nonetheless, the fundamental idea of an aggregate Swiss university is introduced for the first time, and left to prevail the denominational barriers. Balthasar's publication formed one of the starting points later to become important for the Helvetic Society in its development of modern Switzerland. At this time, however, the idea of forming a federal university remained a mere dream.

During the Helvetic Republic (1798-1803), roughly a five-year intermezzo where France attempted to impose a central authority over Switzerland and established a centralized Swiss state based on the ideas of the French Revolution, the idea of a federal university was taken up again—and by the highest authority. Philipp Albert Stapfer (1766-1840), Minister of Education of the Helvetic Republic, suggested in 1799 the establishment of a federal university or a central academy that united the strengths of German universities with the advantages of the Ecole Polytechnique in Paris, which had been established only four years earlier and served as the model for many technical universities. He envisioned that "all essential sciences and arts should be taught in the greatest possible extent and completeness at this ... all-embracing institution". Not only should "attentive and considerate doctors, enlightened philosophers, clear-thinking legislators, competent trustees, expert judges, and sensible scholars" be educated at this institution, but also "creative artist, skilled architects and engineers". Stapfer emphasized that the university in question should be a combination of a university and a polytechnic. It was to be founded not only based on the fact that at the beginning of the 19th century—with the exception of the University of Basel—no actual university existed on Swiss territory, but also on the fact that at that time technical education was limited to a few mediocre "art schools", and everything else was left to private lessons and practice.

With a federal university, Stapfer promised a real national revival of the Swiss people. "This institution will be the focal point of the intellectual forces of our nation, the means to merge the nations still currently separated, compiling cultures from three civilized nations that will form the Helvetic center. It is perhaps destiny to wed

German profoundness with French agility and the refined taste of the Italians."

Stapfer's initiative fizzled out quickly, however. The financial problems of the young state and the impending war were by far optimal conditions in which to move such a project forward. Finally, the short duration of the Helvetic Republic and the decline of Switzerland in the old system of the State Union brought about by Napoleon put the idea to rest.

The restorative phase, which followed from the Congress of Vienna in 1815, did not constitute the appropriate environment to continue the debate on the foundation of a federal university. Only a budding liberalism—Switzerland at that time played a leading role in Europe—was a source of a new dynamic. In his 1827 publication, "Statistica della Svizzera", Stephano Franscini (1796-1857) from the canton of Ticino highlighted the complexities of Swiss reality with a sobering statement on the situation of higher education. Franscini, who was a teacher and an author of textbooks before he became a politician, alleged that Switzerland knows no educational institution comparable to European universities. This led him to call for reform and request that a Swiss university be established. In addition to offering new opportunities to the Swiss academic youth and prosperity to society, Franscini believed that founding a federal educational institution would create a connecting, unifying bond within the Swiss political system.

In 1829, Ignaz Paul Vital Troxler (1780-1866), physician and professor of philosophy, gives the best marks to Swiss compulsory education, claiming it perhaps better than anywhere else in Europe. Troxler's perception of higher education was also rather skewed. The truth of the matter was that Switzerland pathetically took a back seat to Europe! Not a single institution that existed in Switzerland was worth calling a university. Those wanting to continue their education "were forced to leave Switzerland and resort to attending a German university or one in a neighboring country. In this respect, Switzerland was truly far behind."

Even Kasimir Pfyffer (1794-1875), a lawyer and the most important liberal politician in Lucerne, campaigned for the establishment of a Swiss federal university. For him, a university as such was simply a consequence of the liberal movement. In addition to the call for a Swiss Federal Constitution and state, a better education for the people, among other things, was high on the list of priorities. "A federal constitution is to our political life, what a federal university will be to our spirituality. Years will perhaps go by before this comes to fruition, but I have no doubt it will ripen and become reality."

Liberalism significantly progressed the university debate—but only at a cantonal level, not at the federal. The liberal, reformed city cantons particularly benefited. In Zurich, a canton university was founded in 1833; a year later another was initiated in Bern. As a result, the ongoing open debate about a federal university was momentarily laid to rest. Then with the establishment of the Swiss Federal State, the topic resurfaced in 1848 and was immediately drafted into the framework of the constitution (Article 22 of the 1848 Constitution): "The federal government is authorized to establish a university and a polytechnic." What remained a dream and a vision for a good century seemed now a reality—the creation of modern Switzerland was to begin.

The Swiss Federal Constitution of 1848 granted the Confederation the right, but not the obligation to build a university or a polytechnic. However, it quickly became apparent that establishing a national university had not yet gained the political majority, and therefore had no chance. A central university with a range of disciplines in the traditional sense would threaten the existing institutions (Universities of Basel, Zurich and Bern). However, a university limited to technical disciplines would be less competitive.

The three key figures to become the driving force behind materializing a federal polytechnic were Ulrich Ochsenbein, again Stephano Franscini and Alfred Escher. Ulrich Ochsenbein (1811-1890) was the man of power who admitted the university Article into the framework of the Swiss Constitution. As a member of the cantonal government in Bern, as the first president of the National Council and as a Federal Councilor, he understood from the very beginning that instituting a federal university was both a scientific and a national-political task. Language barriers should be torn down and the elite of the young nation brought together. As Ochsenbein saw it, young students were getting their education abroad at the risk of denationalization and returning with ideas and concepts that were not in line with Swiss societal and political conditions.

Stephano Franscini from Ticino, who had campaigned for the establishment of a federal university almost 30 years earlier, and who was now a member of the Federal Council and head of the Department of Home Affairs, was the driving force in gaining the appropriate majority in parliament for at least the establishment of a polytechnic. On 7 February 1854, the bill was passed by Parliament, and on 18 October of the following year the first lectures at the polytechnic took place.

Alfred Escher (1819-1882), a business pioneer from Zurich and member of the State and National Council, was a dedicated advocate to university measures. Just to reach a level of mediocrity among the Swiss universities and academies, Escher asserted, a fair amount of money would be absorbed; but a Swiss national university would be first-rate and could compete with foreign universities. After his hopes for a federal university had finally been crushed, he pushed all the more vehemently for a federal polytechnic and for Zurich as its location, eventually with success. A national university could not have been created, unfortunately, but that was no reason "to ignore the value of the goods gained". This polytechnic was the first of its kind in Switzerland, and thanks to its excellent facilities, the institution became, as Escher claimed, "the center of industrial sciences, particularly important for the industrialization of Switzerland."

*

Nearly 100 years passed between the introduction of the first vague ideas of a national university and the actual founding of the Swiss Federal Institute of Technology, now ETH Zurich. It took several generations of politicians, before the time was ripe. Today's ETH Zurich is the result of this long national debate—the persistent vision to create a federal university was finally realized. But does this mean that the vision has been lived out to the fullest? Inspiring visions often spark new visions— this is evident at ETH Zurich, too. The university is in a constant state of change, as it allows new methods and approaches to science to emerge and evolve. In this light, the Collegium Helveticum was established by ETH Zurich in 1997 to foster collaboration and dialog between the disciplines. Collegium Helveticum, which perceives unites scientists and researchers from ETH Zurich, and since 2004, from the University of Zurich. The institution's core values and vision lie in the exchange between the natural and technological sciences, humanities, arts and medicine. It uniquely exploits the potential of two prestigious universities in Zurich and creates the framework for developing new perspectives in project-related processes of disciplinary exchange. This has clearly demonstrated that dialogue and interaction between the academic disciplines can lead to comprehensive concepts and innovative scientific methods.

Literature
GUGERLI, DAVID, KUPPER, PATRICK, SPEICH, DANIEL. *Transforming the Future. ETH Zurich and the Construction of Modern Switzerland 1855–2005*. Chronos, Zurich 2005.
OECHSLI, WILHELM. *Festschrift zur Feier des fünfzigjährigen Bestehens des Eidg. Polytechnikums mit einer Übersicht seiner Entwicklung*. 2 vol. Huber & Co., Frauenfeld 1905.

Simpler Living in the 2000-Watt Society

OLAF KÜBLER

"Sustainability was yesterday. It's the global economy that matters; if it is supposed to recover we need incessant growth. In times of crisis, you have to worry about the present moment—tomorrow can wait." This seems to be the majority attitude of our current global society.

So much shortsightedness is irresponsible. ETH Zurich and the ETH Domain (École polytechnique fédérale de Lausanne EPFL and four research centers) have not given up the search for viable alternatives towards a fair, sustainable future, especially for the world's richer, developed societies. This is in line with the UN's World Commission on Environment and Development, who in their 1987 report, "Our Common Future" (also known as the Brundtland Report), defined "sustainable development" as such: "Sustainable development is development that meets the needs of the present without compromising the ability of future generations to meet their own needs."

It is no wonder that the environment, sustainability and energy play a prominent role at ETH Zurich. In Switzerland, a land of natural beauty, fully developed with

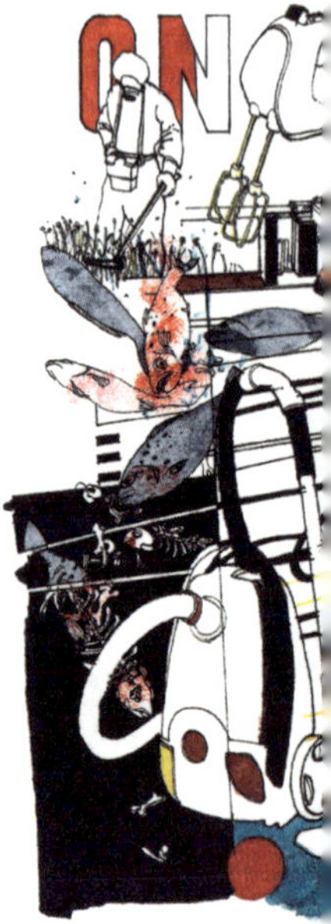

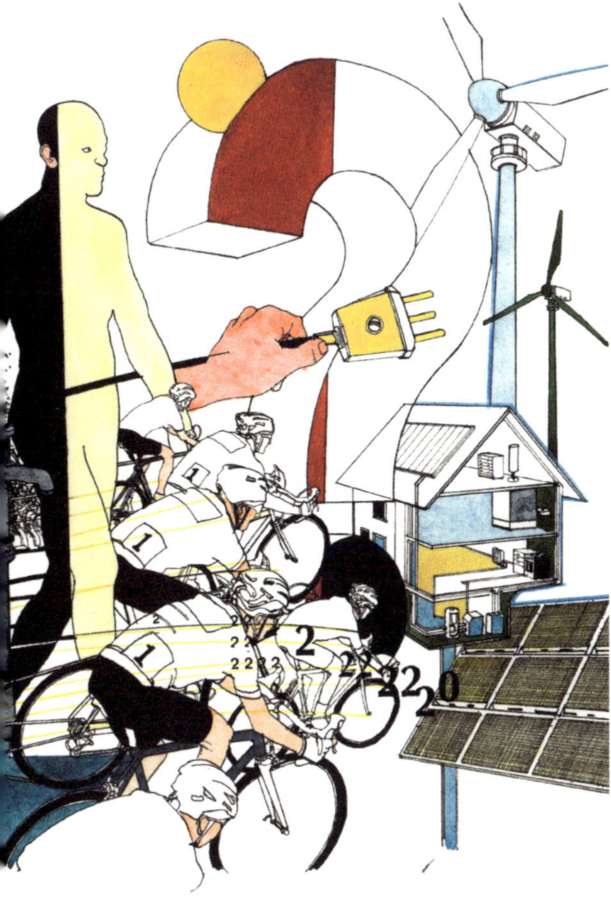

robust growth and prosperity, the threat posed by the destructive clash of technical civilization and the natural environment is highly visible. In this context, we cannot avoid the challenge of taking up ecology, technological development and energy needs as themes of central importance to humankind, nor can we stop working to discover and realize paths towards a sustainable, equitable future.

In the wake of the environmental disasters at Chernobyl (April 1986) and Schweizerhalle (October 1986), ETH Zurich introduced its program in environmental science in 1987, taking the lead in this new field among German-speaking universities. In 1997, the Alliance for Global Sustainability (AGS) was founded in partnership with Tokyo University and the Massachusetts Institute of Technology (MIT), with Chalmers University in Gothenburg joining three years later. Reaching out to urbanized Asia, ETH Zurich established the Singapore-ETH Centre for Global Environmental Sustainability (SEC) in 2010 — in collaboration with the National Research Foundation of Singapore (NRF), the local universities — National University of Singapore (NUS) and Nanyang Technological University (NTU) — and with EPF Lausanne as associated Swiss partner.

The SEC provides the framework for cross-disciplinary collaboration on various research topics. Its first focus has been on "urban sustainability in a global framework." The project has been organized as a Future

Cities Laboratory (FCL), with more than 150 people from over 30 countries on staff and located since 2012 in NRF's brand-new CREATE building (CREATE = Campus for Research Excellence and Technological Enterprise).

Sustainable development, as defined by the World Commission on Environment and Development, is a sympathetic appeal to ethics and justice between generations. But to establish the concrete details of what it involves and then to implement it is difficult from both a conceptual and a practical aspect. If we postulate incessant growth for the world's economy to recover, implementing sustainable development seems next to impossible.

Can continuous growth with its proliferation of modern artefacts of civilization and the concomitant consumption of the environment's resources be at all sustainable? Or is it comparable to the uncontrolled growth of the body's cells at the expense of healthy tissue, i.e. cancer? At best, sustainability and growth can be reconciled if growth is qualitative, not quantitative. Furthermore, it is obvious that qualitative growth must take place within set limits. What limits, then, would be reasonable and practicable?

Our research groups at ETH Zurich, Eawag and PSI were guided by the principle that all human impact on the environment that we could imagine needs to be reversible and need to keep the ecological footprint of humankind in check. In all cases considered, the energy expended—the "work" of society on its environment—turned out to be the controlling factor. This is true for large-scale irrigated agriculture, industrial livestock production or industrial fishing, or when obtaining, selling and using raw materials. It is also true for countless other aspects of our modern technological civilization, such as buildings and infrastructure, motorized transport and tourism—industrial production in general and the many other examples everybody can come up with.

We estimated how much energy is needed to pump one cubic meter of water from an aquifer 1,000 meters under the ground to the surface. The potential energy lies around 3 kilowatt hours, meaning a pump with a 1 kilowatt performance rating would need 3 hours to pump the cubic meter of water. An amateur athlete has a performance of about 2 watts per kilogram of body weight, while a cyclist on steroids such as Lance Armstrong might produce 6 watts per kilogram. A sportsman weighing 100 kilos would therefore need 15 hours to haul the water to the surface. For example, let's consider the Ogallala Aquifer, one of the world's largest aquifers beneath the Great Plains in the United States. It has been used for significant agricultural irrigation since 1950 as electric power and efficient turbine pumps make it cheap to pump the water to the surface. Some current estimates suggest that at the existing rate of usage, the aquifer will run out of water in 25 years.

Or, take an example from your own household: a vacuum cleaner has a performance rating of 1500 watts.

More than three Lance Armstrongs are needed to generate the full suction power of your average home vacuum cleaner. Or how long, and how many slaves did it take to build the great pyramids of Giza? The equivalent volume in soft coal is "devoured" by the largest bucket wheel excavator currently used in Germany in just ten days. The game goes on and on; particularly tempting is to look at how much energy goes into mobility: How many Lance Armstrongs are pedaling under the hood of a premium automobile? A thousand. Or, how many kilowatts were produced by slaves rowing a galley ship?

We estimate that the average performance rating in pre-industrial times was about 300 watts per person, generated by people, draft and pack animals, wind and waterpower. By the year 2000, the performance rating of every human on earth was on average 2 kilowatts. But there are major differences between rich and poor countries here: India was at a mere 0.5 kilowatts per person, Europe at 6 and North America at 13 kilowatts.

Obviously, there are other limits to sustainable development and qualitative growth that can be discussed. The Stockholm Resilience Center has proposed nine research-based "planetary boundaries" or "tipping points": *Climate change* (measured in terms of atmospheric carbon dioxide concentration or increase radiative forcing [W/m^2]) and *Biodiversity loss* (measured in terms: extinction rates) have exceeded the tipping points. For *biogeochemical cycles*, the threshold for anthropogenic nitrogen removed from the atmosphere has been exceeded, while the tipping point for anthropogenic phosphorus going into oceans has not yet been reached. The tipping points have not yet been reached for the dimensions *ocean acidification* (measured in terms of global mean saturation state of aragonite in surface seawater), *freshwater* (measured in terms of global human consumption of water), *land use* (measured in terms of land surface converted to cropland) and *ozone depletion* (measured in terms of stratospheric ozone concentration). There is, as of yet, no agreed way to measure the dimensions *atmospheric aerosols* and *chemical pollution,* and therefore no tipping points have been defined.

The average per capita performance rating is not included in this compilation. It is a heuristic criterion that is easy to understand and measure—and it makes a real difference. But would limits mean sacrificing progress? Certainly, the correlation between energy consumption and prosperity looks so strong that it is treated almost as a natural law. Nevertheless, we venture the hypothesis that it is not a causal relationship. Smart technology and thoughtful use should enable the decorrelation of value creation and energy squandering.

To operationalize this idea, our research group has proposed the concept of the 2000-Watt Society. This is based on the conviction that Switzerland would be able to make do with just a third of today's per capita energy consumption without sacrificing essential amenities of our current lifestyle. The technical expertise for a rich, developed society with low energy consumption and high standard of living seems to be at hand.

Inspired by this principle and convinced by parallel studies and developments at ETH Zurich and its research institutions, the city of Zurich paved the path to realizing the 2000-Watt Society in a 2008 referendum and added the following to its municipal ordinances:

"The community will be actively involved in the protection and preservation of the natural foundations of life and for the sparing use of natural resources. We are committed to implementing sustainable development.

We are responsible for achieving the objectives of the 2000-Watt Society, for a. reducing energy consumption to 2000 watts per capita; b. reducing CO_2 emissions to one ton per person per year; c. promoting energy efficiency and renewable energy sources.

We will no longer invest in or receive power from nuclear power plants."

In the 2008 referendum brochure, the city of Zurich explained how it planned to meet these objectives:

"The city of Zurich is increasingly committed to protecting the environment and conserving natural resources. This amendment to the Municipal Code will anchor environmental concerns in our city's constitution. These goals are based on the 2000-Watt Society, a long-term energy strategy developed by ETH Zurich.

Reducing our energy consumption to a third of its current levels would be achieved through targeted savings, particularly by increasing energy efficiency. By amending the Municipal Code, the city is also committed to a massive promotion of renewable energies.

By the year 2050, energy consumption will have dropped massively and three-quarters of the energy used will come from renewable energy sources. This will make it possible to cut CO_2 emissions in the city of Zurich from today's levels of six tons per inhabitant per year to one ton per year, a level which the climate can better handle. Allowing all existing contracts with the nuclear power plants to expire in the next couple of decades, and forgoing any new investments and contracts in nuclear power will contribute significantly to the cause.

With this new article in the Municipal Code, Zurich establishes itself as the first community in Switzerland to make a constitutional commitment to reduce CO_2 emissions and abandon the use of nuclear energy over the long term."

The commitment to limit per capita energy consumption and its impact on the environment to the world's average usage in 2000 has unleashed some concrete changes. The city of Zurich has already done much to promote energy-efficient buildings and transportation, as indicated in the 2008 referendum brochure:

"How to reduce the city's CO_2 emissions?

The city of Zurich has already taken the first steps needed to reduce CO_2 emissions. Since 2004, the CO_2 emissions have decreased by almost nine percent. In addition, the energy effi-

ciency of buildings has improved in recent years. Old and new buildings today are often better insulated and therefore require less heating. Leading the way are labels for highly energy-efficient building concepts such as 'Minergie' and 'Minergie-P.' In keeping with the 2000-Watt Society, these standards focus on reducing energy consumption while maintaining the same level of comfort. Compared to a conventional new build in 1975, a Minergie-P new building consumes only about 1/10 of the heating energy.

The city of Zurich is developing and already implementing projects related to these goals. The city is requiring the Minergie standard for all new urban construction. Buildings designed according to the more stringent Minergie-P guidelines are also in advanced stages of planning. Examples are the new wing of the Triemli city hospital and the planned Trotte retirement home. Both new buildings even meet the requirements of Minergie-P-Eco and already largely meet the goals of the 2000-Watt Society.

Unfortunately, traffic in the city is not moving in the same direction: However, the amount of traffic in the city is increasing at a significantly slower rate than the Swiss average. This is not only thanks to a well-developed public transport system, but also because a lower percentage of households in the city of Zurich own cars in geographically larger conglomerations.

The city is a walking city. Pedestrians always come first in the city's transport policy. Also important are the concerns of the disabled, the elderly and children. The city is consistently adapting neighborhood centers into pedestrian-friendly zones.

The cycling network is constantly being expanded and the city wants to increase bicycle transportation from seven to twelve percent of overall traffic by 2010.

The role of the 2000-watt limit when building was not yet clear from the proposal submitted in 2008, but the latest planning requirements of the city of Zurich clearly reference it: 'less space, less underground and higher density.' ETH Zurich has contributed to the 2000-Watt and 1 ton-CO_2 Society by developing new heating systems that rescue excess heat, store it in an underground storage tank and have it ready for consumption when heating needs to be increased. ETH Zurich has also been awarded a prize from the International Sustainable Campus Network (ISCN) for its initiative to achieve 100 percent CO_2 free on the Hönggerberg Campus, Science City by 2025."

There remains much to do and different paths have to be taken to create a sustainable energy future. Quantitative concepts for the necessary reductions are: living at home from 1800 to 500 watts, mobility from 1700 to 450 watts, diet from 750 to 250 watts, consumption 750 to 250 watts and infrastructure from 1500 to 550 watts. The hope is that all of this can be achieved under the title of "Simpler Living", as published by Novatlantis, the center for the 2000-Watt Society at ETH (http://www.novatlantis.ch).

The Fascination over New York: "Les gratte-ciel sont plus grands que les architects"

WERNER OECHSLIN

Manhattan pays no attention to what the rest of the world is building, doing or thinking. Nowadays, it can simply lean back and watch the world attempt to replicate the image of New York wherever too much money leads to hype. Yet, a true imitation has never been achieved. One shouldn't be impressed by pictures alone. The skyscrapers of New York stand on rocky ground, which the church founders of Rome couldn't have imagined any better ("tu es Petrus, et super hanc petram aedificabo ecclesiam meam"). By now, Manhattan appears as a stone monument from ancient times, having grown hastily, yet slowly and gradually and long-since blossoming into unmatched uniqueness. As Hugh Ferriss wrote in his book in 1929, it should be the model for the "Metropolis of Tomorrow". What the famous New York Zoning Law sketched out as permissible construction volume ("maximum masses"!), was intended to inspire architects ("here must come architects ..."). For, what was once represented as a loam model could now be or become reality, even with regards to evoking the biblical

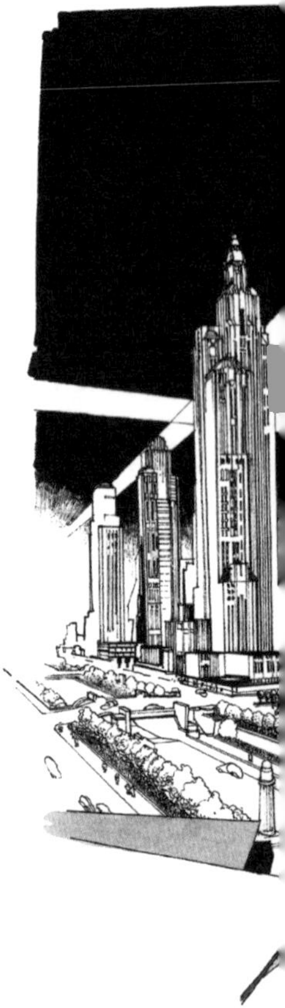

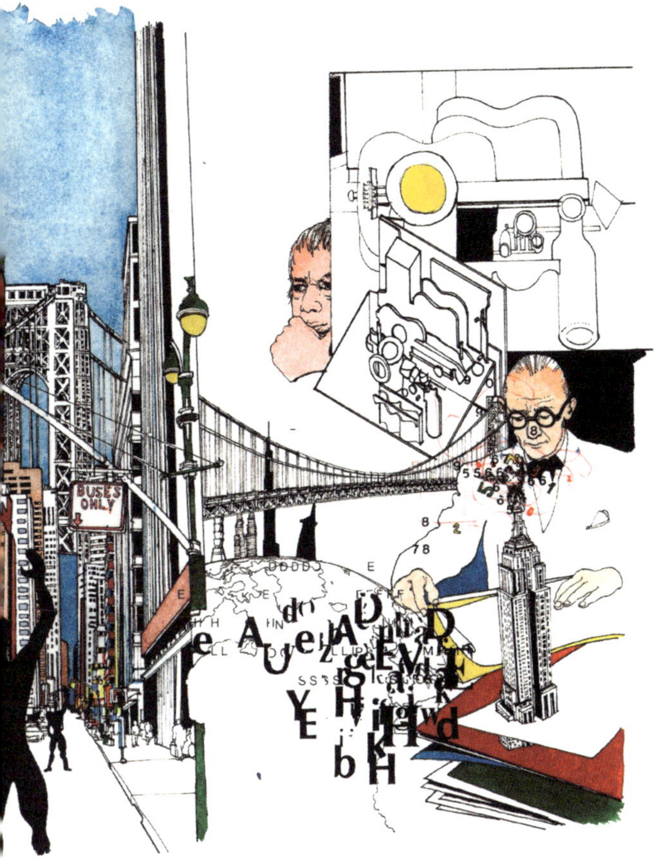
BUSES ONLY

creation myth. "Indeed, the crude clay of the future city may be imagined as already standing." Manhattan epitomizes this archaic architectural drive to build. As Hugh Ferriss pointed out, of the 377 skyscrapers in the USA more than 20 stories high, 188 of them stood "within the narrow limits of New York City". Structures in the Garden of Eden, reaching for the heavens!

The United States has made use of such images and imaginings to generally document its universal claim, its advanced position and function as a role model; as though all of the efforts human society has made toward progress had come together here, for which construction is a visible symbol: "American architecture today is the result of continual world progress—a progress in which building has evolved from centuries of masonry, mass and static concepts to an era of iron, precision and structural dynamics." From the eternal human thirst for knowledge and the interest in the progress and improvement of human living conditions, this was all created on American soil, as is suggested by an advertising brochure from the United States Information Service. The advertisement is illustrated by an image of the UN building planned by Le Corbusier, which the photographer staged as though the vertical, jutting slab was rising above the otherwise naked globe. The globe being the dome of the plenary hall aligns itself with the tradition of images common in Bruno Taut's "Weltbaumeister" and Hugh Ferriss' "Metropolis of Tomorrow". With the statement,

"The most lasting and revealing records of a civilization are its buildings" and a photograph of the Parthenon on the Acropolis in Athens, the brochure opens the hymn and presents the "past and present America" in the right light with an emphatic gesture. American pragmatism has found its quasi-religious counterpart. And this is not new either.

When the publishing house Wasmuth introduced the momentous publication about Frank Lloyd Wright's buildings in Berlin in 1911, a disarming explanation of the exceptional role of American architecture could be read in the preface written by Charles Robert Ashbee: "This is due to the fact that love for the art of building is part of the ethos of the American people." *Omnia vincit amor!* Walther Rathenau saw America as more than just a "Land der unbegrenzten Möglichkeiten", but also as a quintessential "shrewd business" oriented on superlatives, passion and callousness alike that, naturally, evokes envy and wonder. And all of this led Rathenau first and foremost back to the economic facts such as the unfettered course through the global market, whereby "carelessness, waste and interim profit were not hindrances keeping them from driving back goods from other countries". The economic dominance and competition overshadowed everything to which Walther Rathenau stated: "Therefore, America is the country of large margins and an economy of entirety; there is no need to be sparing with such rich sources, and besides, Germany could live off what America wastes." This passage, included in the "Four Nations" essay in 1907, ended with the following: "The extent to which this might pose a risk to the European nations depends on whether these nations will be forced to consider economic development as the measuring stick of the power of a civilization." There is no doubt that "American economic supremacy" is represented in the skyscrapers of Manhattan reaching for the heavens. What can one do to oppose this? America and its pictorial symbol of New York became the standard early on; mixing feelings of admiration with the impression of intimidation. Conflict has since characterized the relationship between Europe and America, especially with regards to architecture and its manifestation, which is overwhelming in Manhattan. Hans Goslar asks in "America 1922": "But where should we take the standard measure from, where should the comparison come from? What right do we have to draw parallels to Europe since a new world was created here that must be given the right [!] to develop autonomously and completely independently in accordance with its own laws?"

*

It is even more amazing that in 1932, modern European architecture, including Switzerland's contribution, experienced its most effective propagation early on, especially in New York, under the successful slogan "International Style". Switzerland was represented by apartment and housing structures and this, not the construction of skyscrapers, is the overall foundation of modern architecture as it was introduced, incidentally, with a focus on the special topic of housing development. The reconstruction of the former exhibition in the then still new Museum of Modern Art by Terence Riley shows that the visitors went straight to the contribution from the offices of Howe & Lescaze where the (only) skyscraper in the exhibition, the Philadelphia Saving Fund Society building was on display. But it wasn't the exhibitions held in the old offices of the Heckscher Building on Fifth Avenue, which were, frankly, very humbly presented, that were responsible for the public impact, but rather it was the catalog and book. The work from the Swiss was thus presented and popularized under the genre "the extent of modern architecture" with photos of buildings from Artaria & Schmidt and the ETH graduate, Max Ernst Haefeli, representing the "Zurich functionalists" and their cohorts.

What has always been viewed as a sort of elective affinity was reconciled, and has been maintained accordingly. The "pure considerations of expediency" emphasized by Ashbee in 1911 as the symbol of American architectural attitude and the recommendation of the Washingtonian advertising brochure "use materials both practical and beautiful" fit perfectly with the Helvetian ideals of frugality and simplicity as Rudolf Schwarz extolled under the motto "Helvetia docet" at the 1948 exhibition on Swiss architecture in Cologne.

However, in 1932, Henry-Russell Hitchcock and Philip Johnson turned their focus to *international style*. Style constituted the category through which the most modern architecture would be annexed and assimilated within history. And, after all, the Helvetian pragmatism was unable to fully satisfy the strict aesthetic ideas still attempted by the elegant, mainly French-influenced city of New York. Haefeli's Züricher Wasserwerkhäuser were presented in photographs with streaked lighting, so the plasterwork, as expressly stated in the caption, gave a contrasting perception against the (modern) flat surface: "rough stucco breaks the effect of surface." What was considered a willingness to compromise from a Zurich perspective, which tended to follow a local tradition, contradicted the aesthetic, formal severity with which modern architecture wanted to adorn itself. This was, after all, a particularly delicate matter. Something pure had to be contrasted with something physical without violating that new ideal. The geometric severity should be expressed in an uncompromising, pure form. Under the principle of "The Avoidance of Applied Decoration" only the smooth surface was compatible. "Simple forms of standardized detail!" No more than that; no raised profile and no surfaces with textures at all. Where such projections constitute merely an unnecessary complication of the wall surface, their effect lacks finishing. In the

1932 catalog of the New York exhibition, this criticism was concretely directed at the houses of Max Ernst Haefeli. The Zurich architectural tradition and pragmatism was thus not very compatible with the "Idea of Style" and the elegance of New York.

*

And yet, at ETH Zurich of all places, a Swiss architect by the name of Bernhard Hoesli, not only propagated "pure modernism" more than anyone and anywhere, he also implemented it into a course of studies, which many correctly say, not only laid the foundation for the Zurich School of Architecture, but also paved the way for the global success of today's Swiss architecture. Naturally, New York was never the sole focus of architectural discussion and study in the USA. Harvard and Cornell stood as standard bearers at the forefront of the modern age and the ETH architect was drawn there more than anywhere else. Alliances were established. Austin, Texas was another location where the "second generation" of modern architecture planned and executed its departure in the early 50^s. Those who were later known as the "Texas Rangers" included architects and artists, among them, Colin Rowe, John Hedjuk, Robert Slutzky, Werner Seligmann and Bernhard Hoesli, who downplayed his prominent role as a leading teacher by referring to his modest origins as a "peasant" from Glarus. However, he was persistent in driving forward architectural form developed from geometry, establishing it into a design method. What was presented at the 1932 New York exhibition and was kept within this context of geometrical form by Hoesli, was translated into a strategy to develop form-giving processes. Hoesli, the ETH graduate who later became an ETH professor, solidified architectural form deriving from geometry, which modern architecture recognized as being originated within the movement de Stijl and its sources, including in that of Hendrik Petrus Berlage, Gottfried Semper's student at ETH Zurich. This led Le Corbusier to say: "Géométrie, seul langage que nous sachions parler." To get to the bottom of the matter, the Texas Rangers believed that one has to start rigorously with the sources, with Mondrian and the (geometric) principles of de Stijl and the creators of the early, exemplary applications by Theo van Doesburg, Cornelis van Eesteren and in the earliest housing structures from Le Corbusier. Hoesli developed his design method on this basis.

As if Le Corbusier's ambiguous position towards the U.S. and New York also encouraged his temporary assistant Hoesli to keep his distance! Fascination alone is deceptive. And finally, New York is a work created by engineers, bridge engineers nota bene, a category to which the very prominent Othmar Ammann, an ETH graduate himself and the creator of the George Washington and Verrazano bridges, belongs. Le Corbusier, who, after all, received an honorary doctorate from ETH Zurich, believed the admiration for what the engineers had achieved was limitless; but with regards to architecture, he stated: "Craignez les architects américains …"

The discussion about skyscrapers was apparently excluded from this development in modern architecture. However, not completely! The European, moralizing discussion stood opposite this development more than in just a critical, rejecting manner. But, as soon as the topic fundamentally moved beyond simple fascination to city and urban development, the tone changed. They wanted to reply to the American new world with the human face of Europe. New York's unique character, those special sounds generated in the narrow streets, evoking strong emotions, first had to be experienced and comprehended in situ. After all, objective America evoked the strongest emotions. The Berlin city planner, Martin Wagner, who, in 1929, at the beginning of a report he drafted in New York about the urban planning problems and their impact on German urban planning, started wondering why hardly anyone was dealing with such matters officially and cited the near sacred words of Emerson: "Build your own world!" Instead, Europe insisted on offering systematically developed solutions. Once again, skyscrapers enter the field of economic dynamics and priority. "Skyscrapers kill" as the old buildings are degraded to "dying buildings" and the increasing property prices lead to unaffordable rents: "The nomadism of the big city to the second power!" For Wagner, New York's narrow streets did not evoke a feeling of urban emotion in any way. He only saw the darkening of the streets on which he commented: "*In the shadow of the titans.* The layperson understands now that rents for the office spaces on the upper floors are more expensive than those on the bottom floors."

There must be a balancing "feeling that guides us through life" and shows us the "factual" way between "philosophical-scientific research" and "poetic enthusiasm", Georg von Buquoy once stated. For Walter Gropius, there are too many traits of a hurried apology. He argued that choosing skyscrapers instead of low buildings "certainly does not require a higher population density". For him, a "psychological" argument of a "known, emotional inclination toward low buildings as the remedy to population density" was in the forefront. The aversion to American design was not architecturally motivated, but rather an idea thoroughly connected to our own habits in life, an idea that continues on here today, where urbanity is occasionally equated with densification, which, in turn, is equated with "bétonnage", or covering everything in concrete.

*

William Dunkel too, born in 1893 in New Jersey, started teaching at ETH Zurich in 1929 after his time in Düsseldorf, attempted to make the matter of skyscrapers more objective and did not condemn the "consistent kitsch" of South American and Spanish skyscrapers. The ten to twelve-story buildings in Buenos Aires and on the Gran Via in Madrid were not offensive in his opinion, despite the "most shameless architectural bareness". And this is particularly true of New York, where so many buildings are coated in art deco patterns. They can be as decorative as they like, those surfaces are subordinate to the great architectural forms. And finally, the matter of skyscrapers is not a stylistic one, but rather one that affects the "economic feasibility of steel frame construction", as Dunkel states in "Das Werk" in 1929.

Those who were familiar with America, having experienced it personally understood this better and were able to handle it. Werner Moser, student of his father, Karl Moser and later a professor at ETH Zurich, experienced it firsthand more than anyone. As René Furer stated in 1978, it was lucky that Werner Moser wrote his America report "in distant Spring Green and not at home"; otherwise, he would have ended up in the argument between rationalists and expressionists and wouldn't have written what was later published in his 1925 essay entitled "Frank Lloyd Wright und amerikanische Architektur" in "Das Werk". Here, he presents Wright's position and the "search for natural expression" as a representation of America: "As a materialist in the actual sense of the word, he recognizes the materials, strives to fits their forms. [...] No preconceived form, no style is important to him beyond this." However, Moser contrasts this position with that which America's architecture actually offers, namely an "image of European art monuments" manifesting everywhere as the "starting point of the form". According to Moser, the Americans thus remain "helpless copyists", which contradicted the "options in the characteristic formulation of the material system that lay in fallow". The skyscraper presents its construction as a steel frame "logically and powerfully" as long as it is free "of the sequined cloak of borrowed styles". In 1925, inspired by Frank Lloyd Wright, this was a clear, unequivocally stated commitment to a modern architecture conceived from the construction and material.

The discussion of moral implications had barely begun at this time, yet their arguments, inversely, lingered for an astoundingly long time. When the "USA builds" exhibition on American architecture opened on 8 September 1945 in Zurich, the Swiss politician and Federal Council member Phillipp Etter gave the opening address and immediately appealed to the "spiritual, mental and Christian attitude" of the American president. Compared to the "great and globally powerful" USA, Switzerland, Etter stated, was by contrast "small, but weak only up against external great power", and yet morally on the same level and willing to be an example of spiritual "réduit". If "the will to build and create new forms" manifests from the exhibitions on American architecture, Switzerland has a moral responsibility to contribute to the "work of the resurrection" in order to overcome "the dynamic, destructive power of the demonic through the static and constructive power of good". On this basis, Phillipp Etter wanted to establish a mutual friendship with the USA. On the other hand, Leland Harrison, representative of the USA, factually held onto that which connected the two countries across the Atlantic, with regards to architecture, and that which came to light in the works of architects and bridge engineers such as William Lescaze and Othmar Ammann, "né à Schaffhouse et diplômé de l'Ecole polytechnique fédérale à Zurich".

The common European prejudice had not only rooted itself in morally conservative politicians, but also in some of the modern architects at ETH Zurich, including Alfred Roth, the loyal follower of Le Corbusier. Roth used that same event to make the assertion that "the human masses and human reason [...] guiding the current unstandardized and random development of the American urban colossi would settle onto a more pleasant path".

American urban colossi! In turn, Werner Moser, who was familiar with the USA, was much more pragmatic and described the American architectural situation as "exceptionally educational". "Because skyscrapers and big cities in America have taken on special characteristic forms", especially in New York, he wanted to look at them in more detail. Instead of lamenting the darkening of the narrow streets, he referred to the "moving light advertisements" that hid the high street walls and created "the illusion of a broad horizon". Darkness and a "confusing abundance of giant neon signs"; that was where he found the "pictorial and romantic aspect" of the big city, which, generations later, has become common in Europe too. Moser saw that which was inevitable due to the tight space, and that what led to structures "towering into the airy space" was a result of a combination of architectural forms with the "development phases of the economy, politics, strategy and, in particular, technical perfection".

This is how Moser dealt with the "American colossal dimension". When comparing New York to Zurich through the lens of local convention and familiarity, and stating that if the Empire State Building was "constructed in Zurich" it would have reached 60 meters below the Üetliberg, Moser most probably evoked shock rather than admiration! In any case, the discussion about large cities and skyscrapers, population density and the American standard still has not gotten of the ground in Zurich 70 years later. However, what has long since become reality in other places, was documented much earlier in the travel notes on America published in 1912 in the *Schweizerische Bauzeitung* by Hendrik Petrus Berlage, who studied at the Polytechnic School with Gottfried Semper: "I had the opportunity of admiring the right principle of the mass impact of these buildings [=residential skyscrapers], in which all adornments, which can only damage this effect, are avoided. Large wall surfaces, in the correct primary distribution of foundation, construction and completion, into which the windows are cut without special framing (the material is brick and thus a pleasant color), must achieve the overall impact of architectural expression—and they do achieve it." The Semper student and later co-founder of the Congrès Internationaux d'Architecture Moderne (CIAM) was fixated on the entire thing and recognized the sign of the times! "I therefore returned from America with the certainty that a new form of architecture was growing there and that this architecture was displaying very significant results." He recognized the "originality and innovation that promises great development in the future".

An early discovery! The statements from Moser that later followed are no different in principle from those of Berlage. Both were based on Frank Lloyd Wright. But the contradictions between America and Europe were often greater and by no means limited along the lines of an advanced America and a limping Europe. They affected the foundation pillars of architecture; the dynamic of material and form and, as Le Corbusier liked to say, of chaos and order. "Je reviens d'U.S.A. Bon! Je vais montrer par l'U.S.A., pris comme exemple, que les temps sont neufs, mais que la maison est inhabitable." This was his argument in 1936 in the *Avertissement* for his work "Quand les cathédrales étaient blanches". It bore the title "Voyage au pays des timides". Only Le Corbusier could afford to sketch an "even better world and idea of a city" than New York symbolized. But Le Corbusier's urban planning recommendations, even those for building on the banks of the Hudson and East River in 1935 remained a utopian idea—and Manhattan is more alive than ever! "AU REVOIR, NEW YORK ... Le bateau long Manhattan. Je sténographe ce paysage émouvant." Le Corbusier, on the other hand, couldn't and did not want to deprive himself of the fascination of New York where its skyscrapers are grander than the architects.

Authors

MICHAEL AMBÜHL
Full Professor at the Department of Management, Technology, and Economics

PETER EDWARDS
Full Professor at the Department of Environmental Systems Science

RALPH EICHLER
President of ETH Zurich

GIOVANNI FELDER
Full Professor at the Department of Mathematics

GERD FOLKERS
Full Professor at the Department of Chemistry and Applied Biosciences, Director of the Collegium Helveticum

LINO GUZZELLA
Full Professor at the Department of Mechanical and Process Engineering, Rector of the ETH Zurich

OLAF KÜBLER
Professor Emeritus at the Department of Information Technology and Electrical Engineering, former President of the ETH Zurich

BEAT H. MEIER
Full Professor at the Department of Chemistry and Applied Biosciences

WERNER OECHSLIN
Professor Emeritus at the Department of Architecture

MARTIN SCHMID
Head of Communications at the Collegium Helveticum

ROLAND SIEGWART
Full Professor of Autonomous Systems, Vice President Research and Corporate Relations of the ETH Zurich

ULRICH W. SUTER
Professor Emeritus at the Department of Materials

GERHARD SCHMITT
Full Professor at the Department of Architecture